Inside MATHEMATICS

Probability & Statistics

HOW MATHEMATICS CAN PREDICT THE FUTURE

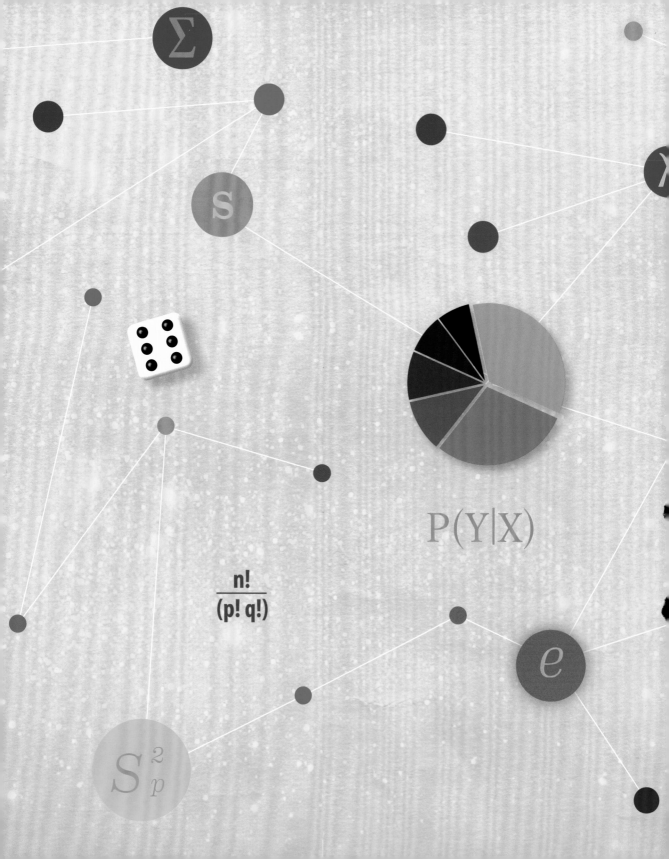

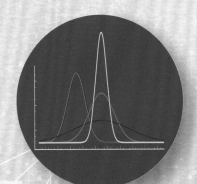

Inside **MATHEMATICS**

Probability & Statistics

HOW MATHEMATICS CAN PREDICT THE FUTURE

Mike Goldsmith

SERIES EDITOR: TOM JACKSON

SHELTER HARBOR PRESS

NEW YORK

4

Introduction

Statistics is the area of mathematics that shows us how to extract meaning from data (usually large amounts of data). A "statistic" is also the name of a piece of information extracted from data, such as an average. We need statistics because so little is certain in life. A lot of things happen by coincidence (chance), and statistics is the tool that tells us all we can know, and what the limits of that knowledge are.

Very often, the things we need to know are described by a whole range of facts. In a school, for instance, there might be a handful of excellent exam results in math, many results in history which are slightly above average, some small classes in which all the students do well, larger classes with problems, and so on. How can this information be summarized in a clear and accurate way? If it can't be, how can the local government decide where help is needed? How can parents choose between this school and another? Statistics helps by defining and calculating the key numbers which describe the schools. Where numbers or facts become too complex to grasp easily, statistics also provides charts to reveal the essentials.

This chart, from the 1850s, shows the effects of food supplements on the growth of pigs. Who said statistics can't be beautiful?

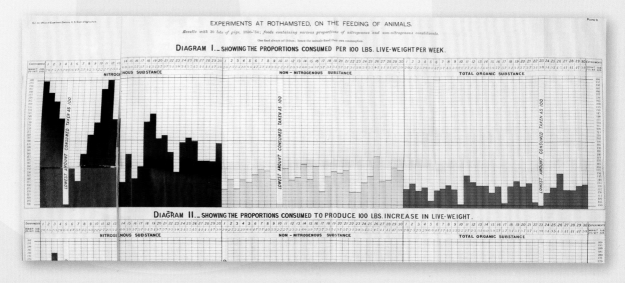

Mount Washington Observatory in New Hampshire has been collecting data on snowfall since 1932. That data tells us that this peak is the snowiest place in the United States.

sixes is about one in a trillion—but how different are these answers? Well, one millionth of an Olympic swimming pool is about 8 cups of water, a billionth is about half a teaspoon, and a trillionth is a droplet a fraction of an inch across.

Small chances

The numbers involved in chances—especially slim ones—can be hard to handle. The chances of throwing 8 sixes in a row with a dice is about one in a million, 16 sixes is about one in a billion, 24

Numerous numbers

Of course, most technical subjects are awash with numbers: the diameters of planets, the loudness of machines, the populations of cities. But there are problems with all these numbers, and it is statistics

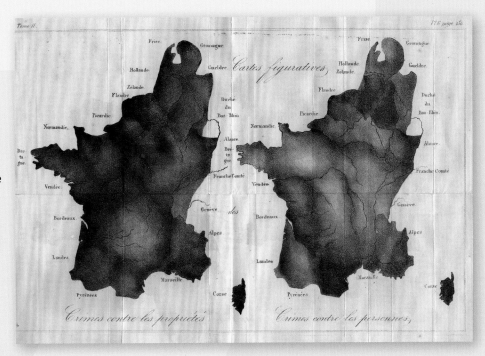

Adolphe Quetelet and his colleagues pioneered a new type of statistical map, in which darkness represents the number of crimes. In these 1831 maps, the left shows crimes against property, the right is crimes against people.

that solves these issues. To begin with, it helps us define our questions precisely. Planetary diameters vary according to where they are measured, noise levels alter constantly, and populations change every moment, as people leave, arrive, are born, and die. So, each question has many answers, and statistics helps us choose, or calculate, the most useful ones. Perhaps the average diameter of the Earth, the maximum noise of a drill, the population of Paris according to the latest census. Another problem is that although in science and math we work with exact numbers, in the real world these are hardly ever known. Our Earth-measuring satellites, noise meters, and censuses all give us numbers, but none can be exactly right. Statistics tells us how to calculate and handle these inaccuracies.

Truth versus hope

Another important job of statistics is to tell us whether the things we believe are true, by removing emotions from our judgments. A researcher with a promising new cure for a disease, an investor excited about the possible future of a new company, or a nervous home owner trying to decide how much to spend on insurance and security, don't have the hard facts they really want. The next best thing is a realistic estimate, but, with so much riding on the answers, realism is hard and optimism or pessimism can take over—unless statistics comes to the rescue.

Probabilities

Of course, statistics can only do so much. It works by collecting data and extracting the maximum information from it, but that may lead to the

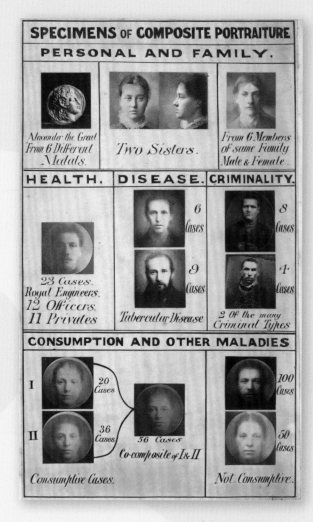

In the 1880s, Francis Galton attempted to use statistical analysis to show that the shape of someone's face could show healthiness and moral character.

conclusion that no conclusion is possible. More often, it will give us probabilities, which can be hard for us to cope with. We want yes/no answers to questions like Am I ill? Will it snow? And, if we don't get those answers, decisions are much harder to take, and we have to take risks instead.

From Frontispiece of Book by WILLIAM PLAYFAIR, An Inquiry Into the Permanent Causes of the Decline and Fall of Powerful and Wealthy Nations, London. 1805.

In 1805, William Playfair attempted to throw light on the long history of trade by using a graphical representaton of trade data in his "Chart of Universal Commercial History." Its message is clear, but is what it says actually true?

Populations and samples

Often, we want to know about things which involve a huge amount of data (like public opinion of politicians, or national health requirements). Ideally, we want to know what every person thinks or needs. In statistics this is known as population data, but it is hardly ever available. Instead, we have to make do with information about a small fraction of the population—a sample. A great deal of statistics is concerned with how to take and analyze samples, and how to predict from them information about the wider population.

Early days

Unlike most areas of math, statistics was not studied in the ancient civilizations of Babylon, Egypt, Greece, India, or China. This is because statistics is often about random chance, and most people in those days thought that the gods determined many aspects of life, natural events as well as possibly the fall of dice in a game. (Even today some gamblers follow superstitions because they think that the score is not down to chance.) This belief in the whims of gods was due to a lack of knowledge about the world. For example, now we know that tomorrow's weather depends on the temperatures, winds, and other conditions today, both here and in distant places, but, in ancient times, none of this was known. The first statisticians were mathematically gifted gamblers, and so statistics was not taken seriously for centuries. Now, often behind the scenes, it rules our lives.

Averages

Aristotle was a fan of averages, or means, but was not overly concerned with mathematical accuracy.

ONE OF THE SIMPLEST STATISTICAL IDEAS IS THE MEAN (OFTEN CALLED THE AVERAGE). AROUND 500 BCE, ARISTOTLE, THE MOST INFLUENTIAL OF ALL THE ANCIENT GREEK THINKERS, TALKED A GREAT DEAL ABOUT THE MEAN. But for him, the word only referred to the avoidance of extremes. He told people they should neither be too brave, nor too cowardly, but aim for something in the middle—the "golden mean."

Aristotle had very little interest in actual data. For example, despite more or less single-handedly inventing the field of biology and being a married man, he incorrectly believed that women have fewer teeth than men. So, Aristotle had no use for statistics. Nevertheless, ancient Greek mathematicians defined (but hardly ever used) a whole series of means, including the geometric mean and harmonic mean. But the only one used today is the arithmetic mean: the sum of a set of values, divided by the number of values in the set.

The unused mean

For centuries after Aristotle, nobody seems to have had the idea of calculating the mean, even though measurements were made for many practical reasons.

For instance by the 1500s, European mariners were starting to explore the oceans. They relied on compasses, which point toward Earth's magnetic pole. This is not the same as the North Pole on maps and globes, so sailors had to measure the difference between the points, known as magnetic declination. Since these measurements were inaccurate, people had to choose from a range of values, but rather than taking the mean, even scholars chose a value near the middle of the range.

Tycho Brahe lost his nose in 1566 in a duel over a mathematical argument.

A world map from 1573 shows straight rhumb lines, or steady courses, for navigating by compass.

Calculating character

One of the first people who calculated and used mean values for magnetic declination was Tycho Brahe, a famous Danish astronomer. (Part of his fame comes from his pet moose, which came to a sad end when it got drunk on beer and fell down a flight of stairs.) But, until about 1800, hardly anyone else used the mean values, and even after that, it was just map-makers and astronomers who used it. Only in the late 19th century did it become a commonplace technique.

What the mean doesn't mean

One reason for this very slow adoption of such a useful and simple statistic seems to be that people were suspicious of a number which was often not seen in the data they had actually collected. This can still be hard to accept today. For example, it may be quite true to say "The average family has 2.2 children and 1.2 cars," but there is no family on Earth which is average in this sense, so the idea is difficult to absorb. The statements mean something very different to "The Smith family has 3 children" or "The typical American family has 2 cars."

Not typical

In fact, what people often want to know is not the mean value, but the typical value. As shown above, a statement such as "The typical family has 2.2 children," is nonsense. The mean can also lead to other strange statements. For example, it is quite correct to say "Most people have more than the average number of legs." Some people have one leg, or no legs, so the mean number of legs is actually slightly less than 2. Since most people have two legs, they have more than the average number. There are also cases where the mean is not the best tool to use. The whole point of the mean is usually to find a middle value, but the mean of bags of potatoes weighing 110 oz, 110 oz, 120 oz, 130 oz, and 230 oz is the not-very-middling value of 140 oz. By the 1920s, statisticians were using two other kinds of average—the median and the mode—(see box, opposite). Together, the trio became known as "measures of central tendency," each best for a particular kind of data.

According to this sample, the average number of feet on the human body is one.

SEE ALSO:
▶ The Normal Distribution, page 46
▶ Outliers, page 108

THE MEDIAN AND MODE

In cases where we have extreme values in our data, a better choice than the mean may be the median, which is the middle observation – 120 oz in the potato case on the page opposite. The median was introduced by Francis Galton (see page 118) in 1881. Like the mean, the median isn't always one of the data: if there are an even number of items, the median is the mean of the middle two. So, the median of 1, 2, 3 and 6 is

$$\frac{(2+3)}{2} = 2.5$$

Another disadvantage of the median is that the data must be sequenced before it can be determined. One problem with both mean and median is that they may not make sense when dealing with whole numbers. If a car salesman wanted to find out how many cars most families have, he might discover that the 10 families he can find out about have 0, 0, 0, 1, 1, 2, 2, 2, 2 and 4 cars. In this case, the mean is 1.4 and the median is 1.5. But the number which occurs most often, which is what the salesperson is most interested in, is 2. This is the mode. It is, roughly, what people mean by the "typical" value. The mode was invented in 1895 by Karl Pearson (see more about him on page 134). Like the median, the mode is easy to work out, but also like the median, the data needs to be put in order first. Also, the mode usually only works with small whole numbers. In a series of accurate measurements, like the weights of 100 humans in grams, or of large numbers, like the populations of all the world's countries, there may well not be any values which are the same, and so there is no mode.

TABLE VIII.—Range in the HEIGHT of Males at each Age and in the several Classes.
(For further details see Tables VIIIa, VIIIb, VIIIc, and VIIId.)

Age in Years	Total number of Observations.	Median Value					Range in Height at each Age									
		Classes				Average of all Classes	Between Upper and Lower Fourths				Average of all Classes	Between Upper and Lower Tenths				Average of all Classes
		1	2	3	4		1	2	3	4		1	2	3	4	
		inches	inches	inches	inches	inches	inches	inches	inches	inches	inches	inches	inches	inches	inches	inches
8–	309	—	—	46·9	47·0	47·0	—	—	3·2	3·4	3·3	—	—	·1	6·1	6·1
9–	514	—	—	49·4	49·1	49·3	—	—	3·0	3·0	3·0	—	—	5·6	6·0	5·8
10–	1533	53·9	52·7	50·9	51·0	52·1	2·7	2·7	3·0	2·9	2·8	5·2	5·3	5·8	6·0	5·6
11–	1766	55·2	53·8	52·3	52·7	53·5	3·1	3·2	3·2	3·7	3·2	6·4	6·3	5·9	5·8	6·1
12–	1960	57·1	55·3	53·6	53·5	54·9	3·4	3·7	3·2	2·7	3·5	6·4	6·6	6·3	7·0	6·6
13–	2743	59·0	57·5	55·3	56·7	57·1	3·8	3·7	2·9	2·7	3·3	7·4	7·2	6·1	5·8	6·5
14–	3419	61·2	59·5	—	59·3	60·0	4·5	4·5	—	3·4	4·1	8·6	8·5	—	6·6	7·9
15–	3497	63·7	62·2	61·9	61·3	62·3	4·5	4·6	2·5	4·0	3·9	8·5	8·6	6·1	7·4	7·4
16–	2780	66·4	65·0	63·6	63·0	64·5	3·7	4·4	2·5	3·7	3·6	7·3	8·5	4·7	7·2	6·9
17–	2745	67·9	66·8	65·8	64·7	66·3	3·5	4·0	2·5	3·2	3·5	6·6	7·1	5·1	6·2	6·3
18–	2305	68·3	67·4	66·4	65·4	66·9	3·4	4·2	3·3	2·9	3·5	6·6	7·1	6·4	5·8	6·5
19–	1434	68·6	67·4	66·5	66·1	67·2	3·3	2·9	3·8	3·1	3·2	6·6	5·7	6·5	5·5	6·3
20–	880	69·1	67·8	67·0	66·5	67·6	3·4	3·5	3·1	2·8	3·3	6·7	6·4	6·5	5·7	6·0
21–	757	68·9	66·9	67·0	66·5	67·3	3·2	2·9	3·4	3·2	3·2	6·5	6·1	6·3	6·4	6·6
22–	516	69·0	67·7	67·2	66·5	67·6	3·7	3·7	3·3	2·9	3·4	7·0	7·2	6·5	5·3	6·2
23–	592	68·5	67·5	67·3	66·2	67·4	2·8	3·2	3·1	3·3	3·1	5·2	6·9	6·5	6·0	5·9
24–	517	68·8	67·2	67·0	66·4	67·1	—	2·4	3·4	2·9	2·9	—	5·3	7·1	5·2	5·8
25–	357	—	67·7	67·4	66·3	67·1	3·0	3·6	2·2	2·9	2·9	—	5·2	6·5	4·7	6·0
26–	315	—	68·0	67·3	66·4	67·2	(3·1)	3·8	3·2	2·3	3·1	(6·0)	6·1	6·5	4·7	5·8
27–	255	(69·4)	68·6	67·6	66·8	67·7	—	3·6	3·2	3·0	3·3	—	6·2	5·6	6·6	6·1
28–	300	—	68·1	67·4	66·6	67·4	(3·1)	3·6	2·8	3·1	3·2	(5·8)	5·2	5·7	5·7	5·7
29–	242	—	68·2	67·4	66·9	67·5	3·2	3·4	3·1	3·2	3·2	5·5	6·5	5·8	6·5	6·5
30–	1010	(69·7)	67·9	67·5	66·7	67·5	(3·2)	2·7	2·4	2·8	3·0	(6·8)	7·0	6·8	6·0	6·5
35–	824	—	68·0	67·6	67·0	67·5	—	3·3	3·5	3·0	3·3	—	6·6	6·5	5·5	6·0
40–	658	(69·0)	68·2	67·3	66·8	67·5	—	2·7	3·3	3·4	3·2	—	5·2	6·7	6·2	6·0
45–	444	—	68·1	67·5	66·3	67·3	—	3·6	3·5	3·4	3·3	—	6·7	8·0	—	
50–	185	—	—	68·2	66·5	67·4	—	—	4·0	3·7	2·6	—	7·7	—	5·2	

NOTE.—The ages under Class I., to which the entries within brackets ( ) apply, were grouped differently to those in the other classes (see Table VIIIa). It has therefore been necessary to exclude those entries from the 'Average of all Classes.'

26

Odds

THE FIRST STATISTICIAN WAS GEROLAMO CARDANO, A MAN AS STRANGE AS HE WAS BRILLIANT. Born in Italy in 1501, he became famous for his medical and astrological skills and traveled through Europe practicing them both— along with many conjuring tricks.

Gerolamo Cardano (right) frequently gave astrological readings for famous people of the time including England's boy king Edward VI.

One of the tricks is still used to this day: an innocent-looking picture book, illustrated in black and white. After an audience member has looked through it, the magician flicks through it again, to reveal the pictures are now in color. It is called a blow book (someone blows on it to start the "magic") and was especially impressive in an age when color picture books were rarely seen.

Bad omens

Perhaps Cardano should have stuck to conjuring. When he cast the horoscope of King Edward VI of England, he was pleased to report that the king would have a long and healthy life. So Edward's death within the month must have been almost as disappointing to Cardano as it no doubt was to the king. He also cast the horoscope of Jesus Christ. To claim that Christ's life could be explained astrologically was a risky endeavor in an intensely Christian society, and despite his famous skills in predicting the future, Cardano did not foresee that he would then be imprisoned in the infamous dungeons of the Italian Inquisition for several months.

Fun with mathematics

At the time, mathematics was a kind of sport, in which mathematicians would challenge each other to solve fiendish problems in mathematical duels. Cardano excelled at this, and loved competitions and games of all kinds—especially games of chance. Most people believed that the outcome of throwing dice or tossing coins was due to the will of God, or could be influenced by magic, or predicted by fortune telling. Cardano realized there was more to it than this, and wrote

Medieval gamblers thought their successes and failures were at the hands of supernatural forces.

a book about the statistics of games of chance. This book was not published in his lifetime, which is perhaps not surprising. For one thing, mathematical skills were closely guarded secrets at the time, since there was money to be made from mathematical tournaments. In addition, the book revealed some of Cardano's favorite methods of cheating. Despite all that, Cardano failed in his plan to make money out of his gambling skills, and concluded that "the greatest advantage in gambling lies in not playing at all."

Dice chances

Cardano was, as far as we know, the first person to apply mathematics to games of chance and to understand that, if you throw a dice, there is an equal probability of getting a 1, 2, 3, 4, 5, or 6, and that there is therefore a one in six chance of throwing a three (or any other number). This chance is also now called the likelihood, probability, or "odds" of throwing a three, and there are many ways to express this (see box).

Fair and balanced

Although this seems obvious to us now, one reason why even a non-superstitious person might have found it hard to grasp in Cardano's time is that a dice which is not quite a cube, or which has one side heavier than another, is not "fair"—that is, it will not obey the above simple

ODDS IN WORDS

When a coin is tossed, there is an equal chance of a head or a tail. Any such equal chance can be expressed in several ways:

1 in 2

0.5

½

50%

"evens"

fifty-fifty

What goes up must come down, but which way will face up?

Strangely enough, there is no single English word to describe this probability. Events which are less than 0.5 probable are referred to as **"unlikely," "possible"** or **"improbable,"** while those with probabilities greater than 0.5 are **"likely"** or **"probable."**

How it works

Want to bet?

In a simple game of chance, an agreed number of dice are thrown at once, and the score is the sum of the results of all the individual dice. Cardano was the first to understand (or at least to write down) how to work out the probabilities of getting various results. Let's say, for instance, that you throw two dice and you place a bet on the total being 10.

The outcome of throwing two dice is a number between 2 (if both dice show a 1) and 12 (if there are 2 sixes). To find out the probability of winning your bet, Cardano realized you can simply list all possible outcomes, and count up the winners.

Even with such a simple game, the possible outcomes make a long list, and an easy way to see them is by a table (though Cardano did not use this method):

		blue dice				
	1	**2**	**3**	**4**	**5**	**6**
1	2	3	4	5	6	7
2	3	4	5	6	7	8
3	4	5	6	7	8	9
4	5	6	7	8	9	10
5	6	7	8	9	10	11
6	7	8	9	10	11	12

(The left-hand axis of the table is labelled **yellow dice**.)

Some numbers occur more than once, because there are different dice combinations which give the same total. If Cardano was betting on the total score being 10, he would count up the number of ways that a 10 can be achieved. As we can see from the table, there are 3, and the total number of combinations is 36. So, the odds of a 10 are 3 in 36, usually abbreviated to "a 3 in 36 chance." Although this is simple enough, one common mistake (and which only a table really makes clear) is to think that there are just two ways to make a 10: either 2 fives, or a six and a four.

rule. Fair dice were not so easy to make in the 16th century as they are now, and deliberately weighted ones were sometimes used to cheat, too.

The will to win

Despite Cardano's insights, he could still not quite believe that there was nothing more to games of chance than mathematics (plus or minus a bit of cheating). He thought that the willpower of the player had an effect on the outcome of games of chance and that "those who throwe timidly are defeated."

Keeping secrets

Cardano loved fame and there's no doubt that he might have become famous as the man who cracked the code of chance. But, thanks to his secrecy, his book remained unpublished until 1663, by which time it was too late to make any difference: his results had been found by others who had already published their findings. In the decades after Cardano, the main thing he was remembered for was probably the time his astrological predictions were completely accurate: he correctly forecast the day of his own death as September 21, 1576. Some say he killed himself to make sure that, for once, he was right.

Independent events

Cardano struggled to work out the probability of getting such results as "2 sixes from 3 throws

In 1532 Cardano reported seeing three suns in the sky above Venice, Italy. They were probably "sun dogs," refractions of sunlight by ice crystals.

of a dice." He gave the correct method at some points in his book, but incorrect methods in other sections. Although there is always the possibility of finding the answer by listing all possible outcomes, and then counting how many give the required result, such lists can be extremely long: 216 lines long for 3 dice throws, over 60 million lines long for 10 throws. And, while tables like the one overleaf work well for two dice, a three-dimensional table would be needed for three dice.

There are formulas for solving such questions quickly, which will turn up in later chapters, but the simplest (though longer) way to work out the answer is this:

The probability of getting a six is

$$\frac{1}{6}$$

and the probability of getting something else is

$$\frac{5}{6}$$

So, we can get the result we want by throwing 2 sixes, then something else. The probability of this is

$$\frac{1}{6} \times \frac{1}{6} \times \frac{5}{6} = \frac{5}{216}$$

We can also get 2 sixes in 3 throws by throwing something else, then 2 sixes:

$$\frac{5}{6} \times \frac{1}{6} \times \frac{1}{6} = \frac{5}{216}$$

Or by getting a six, then something else, then a six:

$$\frac{1}{6} \times \frac{5}{6} \times \frac{1}{6} = \frac{5}{216}$$

There are no other possibilities, so we add up these three to give us our final answer:

$$\frac{5}{216} + \frac{5}{216} + \frac{5}{216} = \frac{15}{216} \approx 0.0694 = 6.94\%$$

This approach is only correct for independent events like dice throwing, in which each throw of a dice does not depend on any previous throws. In games of chance in which, for example, numbered counters are taken from a bag one after another and not put back, the probability changes each time, because there are fewer counters to select from.

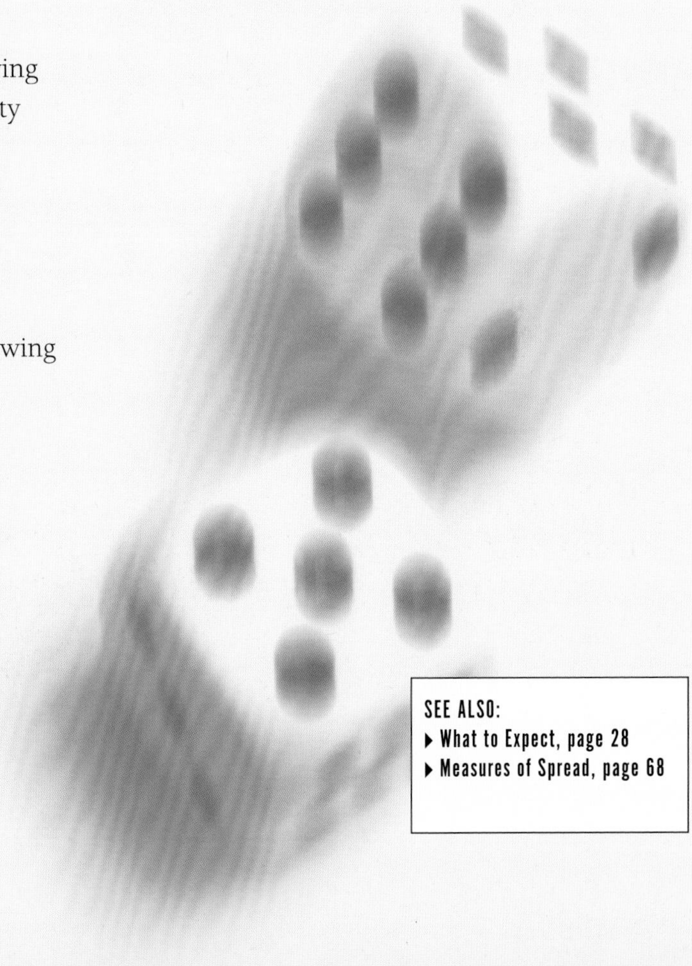

SEE ALSO:
▶ What to Expect, page 28
▶ Measures of Spread, page 68

Combinations and Permutations

DURING THE 17TH CENTURY, INTEREST IN MATHEMATICS GREW AND MATHEMATICAL RESEARCH became a popular hobby for well-off people in Europe with time on their hands. These people were scattered over the continent and, since no-one had yet invented the Internet (nor even the telephone), communicating with each other was not a simple matter.

A French priest called Marin Mersenne made it easier for mathematicians to exchange ideas by setting up an informal "salon," or club, for them. As well as holding meetings, Mersenne also helped them keep in touch by forwarding and copying letters. One of the salon members was a French civil servant called Étienne Pascal.

Marin Mersenne

The temptations of mathematics

Étienne had a son called Blaise, and Étienne made sure Blaise was well educated by the standards of his day—which meant learning languages and literature. Mathematics was not included: Étienne was a kind father who didn't want to overwork his son. However, it wasn't long before Blaise began to study geometry on his own, and enjoyed it so much that his father relented and taught him more. Blaise also started to attend Mersenne's meetings, and soon became known as an excellent mathematician.

The temptations of money

Gambling over games of chance was very popular at the time, and the Chevalier de Méré, a French nobleman who loved to gamble, became friends with Blaise. One day, de Méré asked him about a very practical mathematical problem. In many kinds of gambling, a number of rounds are played, and a score kept of who wins most

Blaise Pascal (second right) is seen here chatting with fellow math genius René Descartes, while behind them Mersenne (third right) is in discussion with Girard Desargues who developed the mathematics of perspective geometry.

rounds. At the beginning of the game, the players all put money into a pot, and at the end of the game, the player who has won most rounds is awarded all the money. The chevalier's question was: what if the game is interrupted when one player is winning? How should the pot be divided?

The Problem of Points

This puzzle, called the Problem of Points, was beyond Pascal, despite having become one of the two greatest mathematicians of his age. Luckily, he was friendly with the other one, another civil servant called Pierre de Fermat, and Pascal asked him for help. Fermat didn't go out any more than he had to, which was just as well since people all over Europe were dying of the plague. In fact, in 1652 it was reported that the disease had killed Fermat too. But the rumors proved false: Fermat survived while many of his senior colleagues perished, which meant that he was swiftly promoted.

The breakthrough is in the post

It wasn't only Fermat who benefited from staying at home; mathematics did too. Because he and Pascal never met, they had to discuss their ideas in letters, which were then available for other salon members to see. Although the pair used different approaches, in the end, both mathematicians came to the same conclusion: the way to solve the problem was by exploring

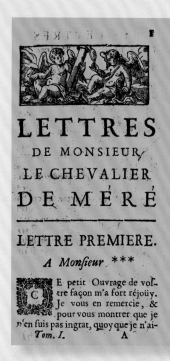

110 *Lettres de Monsieur*

souhaite. de vous une si aimable per-
sonne, & vous l'apprendrez par u
Lettre qu'elle vous en écrit ; M
je vous predis que si vous estes
heureux que de la pouvoir ser
vous me remercierez à quelque heu
de vous en avoir prié. Je vous co
jure donc, Monsieur, de ne vo
pas corriger en cette rencontre d
stre un amy trop violent, & de
croire avec toute l'estime & to
l'affection que je vous dois. Vo
tres-humble & tres-obeïssant se
viteur.

LETTRE XIX.

A Monsieur Pascal.

VOus souvenez-vous de m
voir dit une fois, que v
n'estiez plus si persuadé de l'exc
ence des Mathematiques. V

le Chevalier de Méré. 111

m'ecrivez à cette-heure que je vous
en ay tout-à-fait desabusé, & que je
vous ay découvert des choses que
vous n'eussiez jamais veües si vous
ne m'eussiez connu. Je ne sçay
pourtant, Monsieur, si vous m'estes
si obligé que vous pensez. Il vous
reste encore une habitude que vous
avez prise en cette Science à ne ju-
ger de quoy que ce soit que par vos
demonstrations qui le plus souvent
sont fausses. Ces longs raisonne-
mens tirez de ligne en ligne vous
empeschent d'entrer, d'abord en des
connoissances plus hautes qui ne
trompent jamais. Je vous avertis
aussi que vous perdez par-là un grand
avantage dans le monde, car lors
qu'on a l'esprit vif, & les yeux fins
on remarque à la mine & à l'air des
personnes qu'on voit quantité de
choses qui peuvent beaucoup servir,
& si vous demandiez selon vostre
coûtume à celui qui sçait profiter de
ces sortes d'observations sur quel
principe elles sont fondées, peut-
estre vous diroit-il qu'il n'en sçait

Probability theory began with a letter from a gentleman gambler asking for an edge in games of chance.

all the possible outcomes if the game had not been interrupted. Let's say that the game is won by whichever player gets 3 points, and that it is interrupted when Player A has already got 1 point, while Player B has 0. That means that A needs 2 (or more) points to win, and B needs 3 or more. We imagine the game continuing for four more rounds (because someone will definitely have won by then). There are 16 possible outcomes, and to save space in writing them out we represent Player A winning by an "a" and Player B winning by a "b." So, aaab means that Player A wins the first three rounds and then Player B wins the last one.

The possibilities are:

aaaa, aaab, aaba, aabb, abaa, abab, abba, abbb, baaa, baab, baba, babb, bbaa, bbab, bbba, bbbb.

Remembering that A already has one point, the versions that give Player A two or more points, and therefore an overall win, are these eleven:

aaaa, aaab, aaba, aabb, abaa, abab, abba, baaa, baab, baba, bbaa.

The other five versions give Player B three or more points, and therefore an overall win:

abbb, babb, bbab, bbba, bbbb.

That is, in eleven versions Player A wins, and in five versions Player B wins. So, the answer to the Problem of Points in this example is that Player A is awarded 11/16 of the pot, and Player B gets 5/16.

Math no more

Fermat and Pascal were highly excited by the discoveries they made, but after just a few

Pierre de Fermat is famed for his "Last Theorem," which he never solved. But, he had more success with the theory of probability, for which his contribution is largely forgotten.

count up all the favorable outcomes and divide that number by the total number of outcomes, it was Pascal and Fermat who found ways to do this that did not involve the tedious approach of writing them all down. Like Cardano, Pascal and Fermat only applied their ideas to games of chance, but the tools they developed were later used to solve many other kinds of problems. Their approach works for any situation in which things are arranged in different ways, whether those things are dice, atoms, or human beings.

Factorials

The most useful mathematical tool for tasks involving arrangements is the factorial. The factorial of a number n is written n!, read as "factorial n" and defined as:

$$n! = n \times (n-1) \times (n-2) \times \ldots \times 1$$

Using factorials can be extremely simple. To find out how many ways Anna, Brian, Christine, and Dave can position themselves on a row of four chairs, all we have to do is find 4! (because there are four sitters). And 4! = 4 x 3 x 2 x 1 = 24. We can also use this example to see why this factorial approach works. Anna can sit in any of the four chairs, leaving three chairs free. Brian can sit in any of those three chairs, so together he and Anna have 4 x 3 = 12 options.

months their correspondence ceased. Pascal had found religion, in a very big way, and decided that it would be immoral for him to continue to study either gambling or (rather more surprisingly) mathematics. He did later relax his attitude, but did not correspond with Fermat again. (One of Pascal's final projects was to plan a horse-drawn bus service for Paris; when it was launched in 1662 it was the world's first public transportation system.)

Arranging data

One of the most important legacies of the letters that Pascal and Fermat wrote to each other was their analysis of the ways in which things can be arranged. While Cardano had understood that the most straightforward way to work out the probable outcomes of games of chance was to

There are two empty chairs left, and Christine can sit in either, so between them, Anna, Brian, and Christine have 4 x 3 x 2 options. Dave must sit in the chair that remains, so there is just one option for him, so between them Anna, Brian, Christine, and Dave have 4 x 3 x 2 x 1 options.

Four becomes two

What if only Anna and Brian are involved? How many ways can they sit in the four chairs? To see how this works, imagine for a moment that there were four people as before. Then, as before we would have four choices for the first person, three for the second, two for the third, and one for the fourth. But when there are no third or fourth persons, we stop the calculation after the first two have been seated: four choices for the first person, three choices for the second. Which is 4 x 3 = 12.

As a formula we can write

$$\frac{n!}{(n-r)!}$$

where n is the number of chairs, and r is the number of people. In this case, that is

$$\frac{4!}{(4-2)!} = \frac{4!}{(2)!} = \frac{4 \times 3 \times 2 \times 1}{(2 \times 1)} = \frac{24}{2} = 12$$

exactly as we worked out just now. To see that this is actually the same thing as 4 x 3, look at the third fraction in the equation above. Rather than calculating this, we can instead cancel out all the terms that appear both above and below the line, which gives us our 4 x 3.

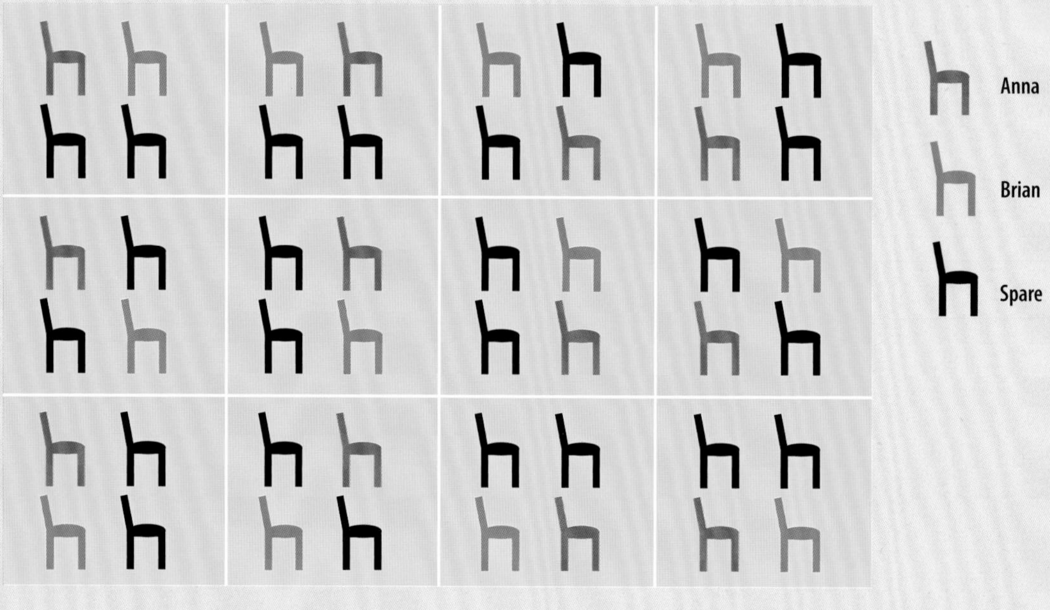

Anna

Brian

Spare

Arranging things of the same type

Sometimes, the items to be arranged are not all different to each other. Let's say we are interested only in the way the males and females in our group of four can choose their seats, but are not interested in which male or female sits where. Listing the possibilities shows that there are only six arrangements: **MMFF, MFMF, MFFM, FMFM, FMMF, FFMM**.

There is a formula for this kind of problem, too, and again it involves factorials. The number of ways of arranging n objects of which p of one type are alike and q of a second type are alike is

$$\frac{n!}{(p!\,q!)}$$

So, the number of ways of arranging 4 people of whom 2 are male, and 2 are female is

$$\frac{4!}{(2!2!)} = \frac{24}{4} = 6$$

Combinations

A combination is an arrangement in which the order doesn't matter. (By contrast a permutation is an arrangement where the precise order does matter.) Let's say you have to buy seven different

Where would you put all the chairs?

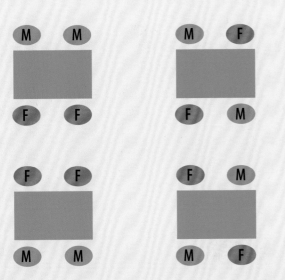

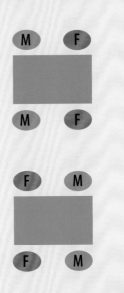

BIGGER PROBLEMS

The point of using a formula rather than listing possibilities is soon clear once slightly larger numbers are involved. Guess in how many ways the letters of the word "aardvark" can be arranged? There are eight letters, so n is 8. "a" is used three times and "r" twice. (d, v, and k appear once and so are not involved in the calculation). So:

$$\frac{8!}{3!2!} = \frac{40,320}{6 \times 2} = 3,360$$

Solving this problem by writing out the possibilities would take nearly an hour, even if you wrote one a second without a break.

seems daunting, the best thing to do is to ask a statistician for help. And their answer would be: just choose the seven meals, don't worry about the order you will eat them in. This will considerably reduce the number of options, and there's a simple rule to find out how large the reduction is. To find this rule without filling the rest of this book with menus, let's just consider three days and three meals: pizza, burger, and pasta. We have these six options in the chart below. If we don't care about the order, these six options turn into just one (burger, pizza, pasta), so we have 1/6 as many options now. Notice here that 1/6 is also 1/3!; the 3! is there because we are considering 3 meals. Going back to our original problem, we have to multiply our answer by 1/number of meals!, which is 1/7!, and that is 1/5040.

meals, one for every day of the coming week (so, in the equation seen earlier, r is 7). You go to the shop and find there are 10 meals to choose from (so, n = 10). The number of choices you have is

$$\frac{10!}{(10-7)!}$$

And the answer to that is 604,800. Lucky you didn't have to write them all down! In case making the right choice out of 604,800 options

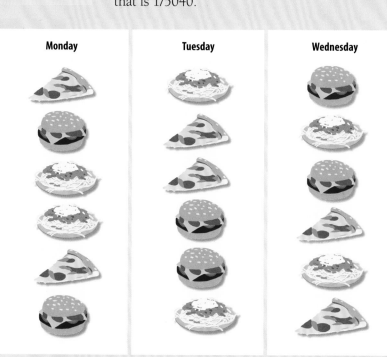

| Monday | Tuesday | Wednesday |

How it works

Combine or permute?

A lottery involves combinations. All we need to win is to get the lucky numbers, it doesn't matter in what sequence, or permutation, they appear on the lottery ticket. But, before you rush off and play a lottery, bear in mind that the numbers involved are still quite large. In a typical lottery, six numbers are chosen from 59. As the sequence doesn't matter, we use

$$\frac{n!}{r!(n-r)!}$$

Putting in the numbers gives

$$\frac{59!}{6!(59-6)!}$$

which is 45,057,474. So, if you buy a lottery ticket once a week for the next 45,057,474 weeks, the chances are you will win. Sadly though, that adds up to just over 866,000 years.
To remember the difference between a combination and permutation, just remember that whoever invented the combination lock, like a safe, wasn't a statistician. In a combination lock, the order of the digits matters. So, it is actually a permutation lock.

So, that gives us

$$\frac{1}{7!} \times \frac{10!}{(10-7)!} = \frac{1}{5040} \times 604,800 = 120$$

So there are 120 options, which isn't quite so horrifying. More generally, when the order matters we use $n!/(n-r)!$, and when it doesn't matter we use $n!/r!(n-r)!$. Different books and sites use different ways of writing $n!/r!(n-r)!$, including $C(n,r)$, ^nC_r, $_nC_r$, and $\binom{n}{r}$. The **C** is for "combination."

SEE ALSO:
▶ The Shapes of Data, page 38
▶ Randomness, page 112

What to Expect

THE MANY LETTERS BETWEEN PIERRE DE FERMAT AND BLAISE PASCAL IN THE EARLY 17TH CENTURY WERE FULL OF NEW IDEAS ABOUT PROBABILITY AND STATISTICS. The letters were copied, exchanged, and discussed by a number of French mathematicians. However, statistics was far from being accepted as a true branch of mathematics, mainly because it was so closely associated with gambling. This started to change in 1655 when Christiaan Huygens, a Dutch scientist, visited Paris and heard about the letters.

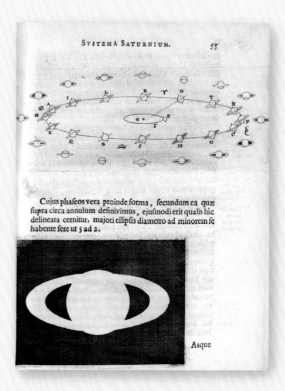

A page from Christiaan Huygens' *Systema Saturnium* of 1659.

Huygens was a man of many interests, including astronomy (he studied Saturn's rings and discovered Titan, Saturn's largest moon), as well as physics (he developed a complete wave theory of light), and timekeeping, where he invented the pendulum clock. Huygens also invented the magic lantern, which is the ancestor of the film projectors used in cinemas. However, he was reluctant to admit that it was his idea, because magic lanterns were so often used to terrify people with frightening images of devils and other monstrous creatures that they soon had a very bad reputation.

Alien monsters

Huygens also tried to use scientific reasoning to work out whether life existed on other planets. He concluded that it did. Several other people at around this time were also suggesting this, but while they often assumed intelligent living things elsewhere would look very similar to us, Huygens realized that the aliens, or Planetarians, as he called them, might not. They might have "Flesh on the inside of their Bones" and "Great round saucer Eyes five or six times as big [as ours]." Despite being one of the great inventors of his time, he

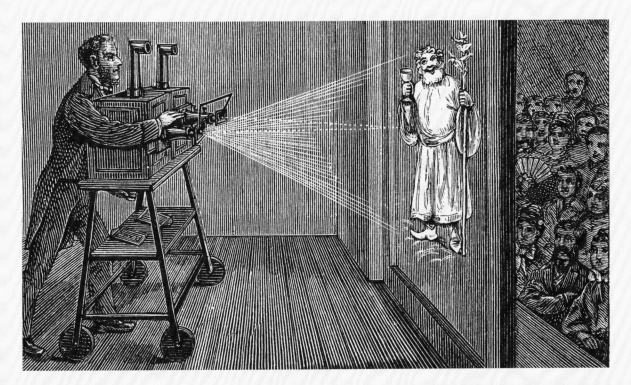

Above: The magic lantern was an early projector and was used to create a ghostly special effect during theater performances.

Right: A Greek edition of Huygens' masterpiece *Cosmotheoros*.

unfortunately did not make much progress in working out how to get to other planets to find out if he was right, only saying that "some Pegasus" (a flying horse from Greek mythology) might help.

First in print

Pascal (see more, page 20) was so impressed with Huygens' skill that he pestered the Dutchman into writing a book about statistics, which Huygens did. *De ratiociniis in ludo aleae* (On Reasoning in Games of Chance) was first published in 1657. It was the first published book about what we now call statistics, and remained the most popular book on the topic for at least half a century.

CHRISTIANI
HUGENII
ΚΟΣΜΟΘΕΩΡΟΣ,
SIVE
De Terris Cœleſtibus, earumque ornatu,
CONJECTURÆ.
AD
CONSTANTINUM HUGENIUM,
Fratrem:
GULIELMO III. MAGNÆ BRITANNIÆ REGI,
A SECRETIS.

ACAD.
LUGD BAT
BIBL.

HAGÆ-COMITUM,
Apud ADRIANUM MOETJENS, Bibliopolam.
M. DC. XCVIII.

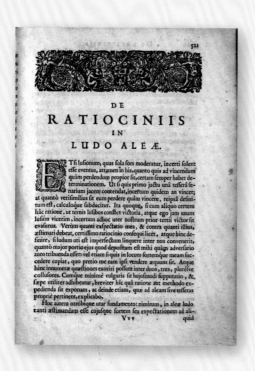

The title page of *De ratiociniis in ludo aleae*, Huygens' 1657 book on games of chance.

Great expectations

Huygens also took a key step toward turning statistics into a powerful tool for analyzing data of all kinds by introducing the concept of expectation. Like many statistical ideas, expectation now seems obvious. It is what you expect to happen when you carry out some random process many times. This does explain clearly what we mean when we say that there is a 1 in 6 chance of getting a two (let's say) in throwing a dice. If you throw a dice once, you would have no more reason to expect a two than any other score, but throw it 6,000 times, and you would expect to see about 1,000 twos.

QUANTUM STATISTICS

Radioactive materials, like radium, release their light and other forms of energy according to random processes. Each of these processes has an expectation value, and these values are worked out by applying statistics to quantum physics.

Expected averages

A slightly less obvious example is this: what average score would you expect when you throw a dice many times? Before reading on, try guessing the answer. If you threw a dice 1,000 times, adding the scores each time, and then divided the

Dice and gambling are ancient. The oldest dice is almost 5,000 years old.

How it works

Is it worth it?

Expectation values are often used in making business decisions about whether a plan is worthwhile pursuing. For example, the manager of a company trying to make gadgets as cheaply as possible knows that about 1 in 100 will be faulty, but cannot afford to test them all. Faulty ones will be returned for a refund. The manager can only undercut the competition if the gadgets are priced at $25 each. This includes a profit of $1. Is it worthwhile selling the gadgets? We work out the expected value (*E*) like this:

$$E(X) = (\$1 \times \frac{99}{100}) - (\$25 \times \frac{1}{100}) = \$0.99 - \$0.25 = \$0.74$$

The $0.74 is a profit, so this is a worthwhile plan. But, as always with expectation values, this is the profit per item to be expected on average, once many items have been sold. So, the plan will only be a success in the long run. If there is 1 faulty gadget in the first 10 sold, the company will make a loss of

$$\$1 \times 9 - \$25 \times 1 = -\$15$$

total by 1,000, what value would you expect? To find out, we write the probability of every possible outcome (*X*). For instance, the probability (*P*) of throwing a 4 is 1 in 6, which we can write as

$$P(X=4) = \frac{1}{6}$$

We can write this for every possible outcome:

$$P(X=1) = \frac{1}{6} \qquad P(X=2) = \frac{1}{6}$$

$$P(X=3) = \frac{1}{6} \qquad P(X=4) = \frac{1}{6}$$

$$P(X=5) = \frac{1}{6} \qquad P(X=6) = \frac{1}{6}$$

These are our probabilities. Now we turn them into scores. As this is a very simple game, this is a

very easy step. If we throw a 1, we get a score of 1, and so on. Applying this to every outcome, we get these expectation values (*E*):

$$E(X) = 1 \times P(X=1) + 2 \times P(X=2) + 3 \times P(X=3)$$
$$+ 4 \times P(X=4) + 5 \times P(X=5) + P(X=6)$$

And finally, we use the actual values for these scores that we calculated earlier and add them up:

$$E(X) = \frac{1}{6} + \frac{2}{6} + \frac{3}{6} + \frac{4}{6} + \frac{5}{6} + \frac{6}{6} = \frac{21}{6} = 3.5$$

So the average score from many dice throws is 3.5.

SEE ALSO:
▸ Averages, page 10
▸ Measuring Confidence, page 154

Matters of Life and Death

IN THE 1660s, LONDONERS WERE DYING IN THEIR HUNDREDS EVERY DAY, as plague raged through the city. With no understanding of germs, no one knew how to cure it. The only thing people could do was to move away from those areas where the plague was the most prevalent.

They knew where those areas were because, since 1611, once a week the Worshipful Company of Parish Clerks had been producing a "Bill of Mortality," listing the numbers of people who had died in different parts of London, and what had killed them.

Boiling down the data

John Graunt, a London merchant, realized that a lot more could be found if the information from different weeks was pooled. So, he began to analyze and list the data, and in doing so invented demography, the statistics of human populations. Insurers, governments, health services, and many others rely on demographics today. Graunt did more than just adding up the data, he evaluated it, too, realizing that, for example, when very old people were recorded as dying of coughs, it would be safer to say they had died of old age. Graunt published his findings in 1662, as *Natural and Political*

London 19 From the 25 of April to the 2 of May. 1665

Parish	Bur.	Plag.	Parish	Bur.	Plag.	Parish	Bur.	Plag.
St Alban Woodstreet			St George Botolphlane			St Martin Ludgate		
Alhallows Barking	3		St Gregory by St Pauls	3		St Martin Orgars		
Alhallows Breadstreet			St Hellen			St Martin Outwich	1	
Alhallows Great	1		St James Dukes place	1		St Martin Vintrey	3	
Alhallows Honylane			St James Garlickhithe	1		St Matthew Fridaystreet		
Alhallows Leffe	1		St John Baptift			St Maudlin Milkstreet		
Alhallows Lumbardstreet			St John Evangelift			St Maudlin Oldfishstreet		
Alhallows Staining	2		St John Zachary			St Michael Baffishaw		
Alhallows the Wall	3		St Katharine Coleman	2		St Michael Cornhil		
St Alphage			St Katharine Crechurch	1		St Michael Crookedlane		
St Andrew Hubbard	1		St Lawrence Jewry			St Michael Queenhithe		
St Andrew Underhaft			St Lawrence Pountney	1		St Michael Quern		
St Andrew Wardrobe	5		St Leonard Eastcheap			St Michael Royal		
St Ann Aldersgate	2		St Leonard Fosterlane	1		St Michael Woodstreet	1	
St Ann Blackfryers	1		St Magnus Parish	2		St Mildred Breadstreet	1	
St Ancholins Parish			St Margaret Lothbury			St Mildred Poultrey	2	
St Austins Parish			St Margaret Moses			St Nicholas Acons		
St BartholomewExchange			St MargaretNewfishstreet	1		St Nicholas Coleabby	1	
St Bennet Fynck	1		St Margaret Pattons			St Nicholas Olaves		
St Bennet Gracechurch			St Mary Abchurch			St Olave Hartstreet		
St Bennet Paulswharf			St Mary Aldermanbury			St Olave Jewry	3	
St Bennet Sherehog	1		St Mary Aldermary			St Olave Silverstreet		
St Botolph Billingsgate			St Mary le Bow			St Pancras Soperlane		
Chrifts Church	1		St Mary Bothaw			St Peter Cheap		
St Christophers			St Mary Colechurch			St Peter Cornhil	1	
St Clement Eastcheap			St Mary Hill			St Peter Paulswharf		
St Dionis Backchurch	1		St Mary Mounthaw	1		St Peter Poor		
St Dunstan East	2		St Mary Sommerset	2		St Steven Colemanstreet	2	
St Edmund Lumbardstr.	1		St Mary Stayning			St Steven Walbrook		
St Echelborough	1		St Mary Woolchurch	2		St Swithin	1	
St Faith			St Mary Woolnoth			St Thomas Apostle	1	
St Foster	1		St Martin Iremongerlane			Trinity Parish	1	
St Gabriel Fenchurch	2							

Buried in the 97 Parishes within the Walls — 70   Plague — 0

Parish	Bur.	Plag.	Parish	Bur.	Plag.	Parish	Bur.	Plag.
St Andrew Holborn	14		St Botolph Aldgate	8		Saviours Southwark	16	
St Bartholomew Great	4		St Botolph Bishopsgate	11		S. Sepulchres Parish	13	
St Bartholomew Lesse			St Dunstan West	6		St Thomas Southwark		
St Bridget	8		St George Southwark	5		Trinity Minories		
Bridewel Piecinct			St Giles Cripplegate	18		At the Pesthouse		
St Botolph Aldersgate	4		St Olave Southwark	16				

Buried in the 16 Parishes without the Walls, and at the Pesthouse — 125   Plague — 0

Parish	Bur.	Plag.	Parish	Bur.	Plag.	Parish	Bur.	Plag.
St Giles in the fields	24		Lambeth Parish	5		St Mary Illington	3	
Hackney Parish			St Leonard Shoreditch	8		St Mary Whitechappel	11	
St James Clarkenwel	13		St Magdalen Bermondsey	15		Rotherith Parish	2	
St Kath. near the Tower	5		St Mary Newington	4		Stepney Parish	36	

Buried in the 12 out Parishes in Middlesex and Surry — 127   Plague — 0

Parish	Bur.	Plag.	Parish	Bur.	Plag.	Parish	Bur.	Plag.
St Clement Danes	13		St Martin in the fields	26		St Margaret Westminster	20	
St Paul Covent Garden	5		St Mary Savoy	2		whereof at the Pesthouse		

Buried in the 5 Parishes in the City and Liberties of Westminster — 66   Plague — 0

B 3

This *Bill of Mortality* from 1665 lists London's dead.

The Diseases, and Casualties this year being 1632.

Abortive, and Stilborn	445	Jaundies	43
Affrighted	1	Jawfaln	8
Aged	628	Impostume	74
Ague	43	Kil'd by several accidents	46
Apoplex, and Meagrom	17	King's Evil	38
Bit with a mad dog	1	Lethargie	2
Bleeding	3	Livergrown	87
Bloody flux, scowring, and flux	348	Lunatique	5
Brused, Issues, sores, and ulcers,	28	Made away themselves	15
Burnt, and Scalded	5	Measles	80
Burst, and Rupture	9	Murthered	7
Cancer, and Wolf	10	Over-laid, and starved at nurse	7
Canker	1	Palsie	25
Childbed	171	Piles	1
Chrisomes, and Infants	2268	Plague	8
Cold, and Cough	55	Planet	13
Colick, Stone, and Strangury	56	Pleurisie, and Spleen	36
Consumption	1797	Purples, and spotted Feaver	38
Convulsion	241	Quinsie	7
Cut of the Stone	5	Rising of the Lights	98
Dead in the street, and starved	6	Sciatica	1
Dropsie, and Swelling	267	Scurvey, and Itch	9
Drowned	34	Suddenly	62
Executed, and prest to death	18	Surfet	86
Falling Sickness	7	Swine Pox	6
Fever	1108	Teeth	470
Fistula	13	Thrush, and Sore mouth	40
Flocks, and small Pox	531	Tympany	13
French Pox	12	Tissick	34
Gangrene	5	Vomiting	1
Gout	4	Worms	27
Grief	11		

Christened { Males—4994, Females—4590, In all—9584 } Buried { Males—4932, Females—4603, In all—9535 } Whereof the Plague—8

Increased in the Burials in the 122 Parishes, and at the Pesthouse this year—993
Decreased of the Plague in the 122 Parishes, and at the Pesthouse this year—266

John Graunt (right) used the raw data to show the frequency of different causes of death.

Observations ... Made upon the Bills of Mortality. The many discoveries Graunt made through his analysis showed just how useful statistics could be. He found that more boys were born than girls, but that more males died per year than females, which led to roughly similar numbers of adult males and females in London. He also proved that the popular idea that plagues appear when a new king is crowned was nonsense.

Scientific approval

The importance of Graunt's book was recognized at once, and the same year he was elected to the Royal Society, a scientific club almost entirely populated by much wealthier men (women members were forbidden). As a working man, Graunt would not have been welcome without the support of the king, Charles II, who was probably pleased to hear Graunt's proof that his accession hadn't caused the plague. Charles even directed the society "if they found any more such Tradesmen, they should be sure to admit them all, without any more ado."

In 1666, Graunt's business was destroyed in the Great Fire of London and, to add insult to injury, he was accused of starting the fire and helping it spread. Although he was exonerated, and found employment as a manager of one of London's water supply companies, he had financial problems for the rest of his life.

SEE ALSO:
▶ Bayes' Amazing Theorem, page 52
▶ Seeing Statistics, page 64

In the 1660s the plague was thought to spread by air, so doctors wore masks packed with herbs to filter out the disease.

How Long Will You Live?

EDMOND HALLEY IS MOST FAMOUS TODAY AS THE MAN WHO CORRECTLY PREDICTED THE RETURN OF THE COMET WHICH NOW BEARS HIS NAME. However, he had many other interests. He invented and used a new kind of diving bell, and traveled the world studying geology, astronomy, weather systems, and Earth's magnetism.

In a period full of bad-tempered scientists, Halley was also well known for his charm. He used it on Isaac Newton, one of the most bad-tempered

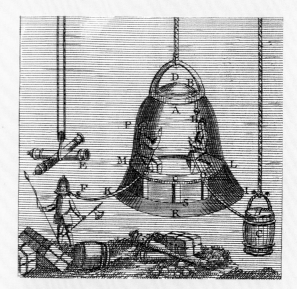

Above: Diving bells offered a way of exploring the hidden world of the ocean floor.

scientists of them all, when Edmond managed to convince Isaac to publish *Philosophiae naturalis principia mathematica*, (Mathematical Principles of Natural Philosophy), which explained the laws of motion and gravity and was probably the most important science book ever written. (Halley ended up funding the publication in 1687.) King Charles II liked Halley so much

Sir Edmond Halley.

City of changes

In statistics, Halley's small but significant contribution involved the application of statistics to the length of human life. Halley was inspired to do this by the work of John Graunt (see page 32), but when he studied Graunt's data about London he found something that initially seemed very odd. On the one hand, there were many more deaths than births (largely due to plague), but on the other hand, the number of people

that he ordered Oxford University to give him a degree. It may have helped that Halley named a new constellation in the sky of the Southern Hemisphere after the oak tree in which Charles hid after being defeated in battle.

Above: A headstone in a London cemetery marks the grave of a victim of the Great Plague of the 1660s.

Left: A map of London prior to the Great Fire of 1666.

Far left: The future King Charles II famously had to hide in an oak tree as he fled from murderous rebels during the English Civil Wars in the 1640s. His father Charles I was not so lucky and was executed.

Left: The old city of Breslau in Poland, now named Wroclaw.

1,238 babies born in Breslau

reached 7 years

died within 2 months

in London was steadily increasing. The only possible explanation was that many people were moving into the city from the country.

Mortality tables

Halley saw that this was a problem. What he wanted was effectively to track the lives of individuals over many years to establish reliable "mortality tables," which would answer questions like "what is the probability that a 50-year-old Londoner will live another 10 years?" Such questions are essential today to insurance companies, and they come down to "how many years is a baby born in year X likely to live?"

The Breslau solution

Halley worried that, because so many Londoners had not been born in London, it was not possible to draw conclusions about their whole lives (and Graunt's data was only available for London). So, he chose a city which had records

as good as London's, but with a more stable population—the city of Breslau in Poland. Using this data, Halley went on to develop methods of calculating insurance rates. As well as its statistical importance, Halley's Breslau analysis reveals some sad facts about life in those times. Of 1,238 babies born per year, 348 died within 2 months (today, we would express this as a 28 percent first-year mortality rate). And, just 692 of those 1,238 reached their seventh birthdays.

Changing rates

Since Halley's time, mortality rates have remained of great interest, but it has become popular to talk about survival rates, too, even though these rates can be misleading (see box, right). It is more important than ever to be clear about these statistics, because medical science is now able to change death rates and save people's lives by government investment in selected treatments. Since money is limited, governments must choose the best treatments to spend their money on, and "best" can refer either to highest survival rate, or to lowest mortality rate.

How it works

Mortality versus survival

The mortality rate of a particular disease is the percentage of population who die of that disease in a year. For instance, the mortality rate for people with lung cancer in the United States is 0.0534 percent, which is more often expressed as 53.4 deaths per 100,000 people. The survival rate is the percentage of people with the disease who are still alive a certain number of years (five years is often used) after diagnosis. The five-year survival rate for people with lung cancer in the USA is 15.6 percent. If a cure for a disease is found, or a way to extend the lives of people who have that disease is discovered, the survival rate will rise and the mortality rate will fall.

But, of course, there are many diseases which cannot be cured. Let's say there is a disease called ABC. Every day it kills about 200 people somewhere on Earth (which is about 75,000 per year). That means that ABC kills 1 in 100,000 people per year; that is a mortality rate of 0.001 percent. If it is fatal within 3 years of the first symptoms appearing, that means its 5-year mortality rate is 100 percent. Now, usually, medical tests can detect the presence of an incurable disease long before any symptoms appear. If, in the case of ABC, a new test was developed which detected the disease 3 years before the first symptoms appeared, that would change the 5-year mortality rate from 100 percent to 0 percent (since people would die 6 years after diagnosis), but no one's life would actually have been saved. Although this is an invented example, there is much more progress in diagnosing diseases than curing them, and, because politicians like good numbers, it is tempting for them to put lots of money into early diagnoses, because that can have such a big effect on survival rates. Of course, early diagnosis can often lead to better treatments, but not always, and it is important that the limited funds available are spent both on researching new cures and treatments as well as on earlier diagnoses.

What are the chances?

SEE ALSO:
▶ Matters of Life and Death, page 32
▶ The Average Human, page 96

The Shapes of Data

Jacob Bernoulli.

Although Cardano, Fermat, Pascal, and Huygens had developed many statistical tools and formulas, they had used them only to study games of chance. The person who did most to extend the reach of this powerful new kind of mathematics was Swiss mathematician Jacob Bernoulli.

Jacob (seen above) and his brother Johann were both brilliant mathematicians (and so were Johann's three sons and two of his grandsons). All this mathematical genius might have been wasted if Jacob and Johann had listened to their father, who thought that they should do something sensible, like taking over the family spice trading business, or becoming priests or physicians. Jacob reluctantly attended courses in theology at university, but studied mathematics and astronomy there, too. After obtaining his

Known as a bean machine, Galton board, or quincunx, this device creates a physical representation of data distributed by chance.

theology degree in 1676, he did not go back to his home city in Basel, which is now in Switzerland, but instead traveled widely, meeting many mathematicians in France, the Netherlands, and England.

Johann versus Jacob

He finally returned home in 1683, where he taught mathematics and physics at Basel University; in 1687, he became a professor of mathematics there. As was common at the time, he also chose a motto. His came from Phaethon, the son of the Sun god of ancient Greece. Phaethon disobeyed his father and took control of the chariot that drew the Sun across the sky, and Jacob's motto was "Invito patre sidera verso" which means "Despite my father, I am among the stars." Meanwhile, Jacob's brother Johann had been doing what his father wanted and was

The University of Basel was founded in 1460 by Pope Pius II and is one of the oldest colleges in the world.

Johann and Jacob Bernoulli discuss geometrical problems.

Johann Bernoulli.

studying medicine at Basel, but also followed his brother's example by studying mathematics. The brothers worked together for a while to develop the mathematics of calculus, a powerful new tool for the study of changing quantities which had recently been discovered by Isaac Newton in England and Gottfried Leibniz in what is now Germany. Jacob was soon to apply calculus to probability, but fell out with his brother even sooner. Over the next few years, Johann described Jacob as "verbose, ambitious, greedy, secretive, misanthropic, envious, and proud."

The law of large numbers

When he wasn't insulting his brother, Johann was struggling to answer a simple yet fundamental question: what does it really mean to say that, if you toss a coin, it is as likely to be a head or a tail? Let's imagine tossing a coin and it falling as a head. How is this different to a ball falling to the ground rather than rising up to the ceiling? You might say that there was only a 50 percent chance of the head, but a 100 percent chance of the ball falling, but how do you know that? To find out, you could repeat the experiment. Toss the coin, and drop the ball again. The ball falls each time, but with the coin, you might easily get another head. What Bernoulli saw is that the secret to the puzzle is to repeat the experiment not just once or twice, but many times. In fact, when he

Jacob Bernoulli's work on probability was contained in his 1713 book *Ars conjectandi* (The Art of Guessing).

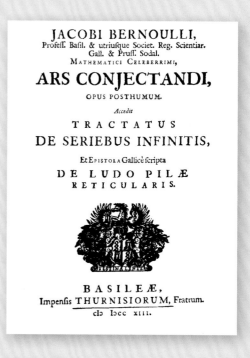

JACOBI BERNOULLI,
Profess. Basil. & utriusque Societ. Reg. Scientiar.
Gall. & Pruss. Sodal.
MATHEMATICI CELEBERRIMI,

ARS CONJECTANDI,

OPUS POSTHUMUM.

Accedit

TRACTATUS
DE SERIEBUS INFINITIS,

Et EPISTOLA Gallicè scripta
DE LUDO PILÆ
RETICULARIS.

BASILEÆ,
Impensis THURNISIORUM, Fratrum.
clɔ lɔcc xiii.

formulated this as a mathematical law, it became known as the law of large numbers. In words, it says that the more often you repeat a statistical experiment, the more closely its outcome will match its probability.

The real world

Despite its apparent obviousness, this law was a breakthrough, because it meant that at last the mathematics of statistics could properly tackle the real world. It could be applied to all random events, not just coin tossing, and it could be rigorously proved, too. But proving the law was at the very limit of what mathematics could do. Even Jacob Bernoulli, by now probably the greatest mathematician of his day, struggled for 20 years to do so.

Hard questions, easy answers

It's simple to answer the question "What is the probability of getting a head if you toss a coin?" But, what about "What is the probability of getting one head and one tail if you toss two coins?" The easiest way to answer this is to follow the method used by Gerolamo Cardano:

1. List the possible outcomes (HH, HT, TH, TT).
2. Count them (4).
3. List how many of those outcomes you are interested in (TH, HT).
4. Count those (2).
5. Express that result (2) as a fraction of the total (4), to give 2/4, which is ½, or 0.5.

The same approach can be applied to more complicated questions, too, like "What is the probability of getting five heads in seven throws?" or "What is the chance of getting a total of 7 by throwing four dice?" but it gets increasingly long-winded. If only there was some way of converting these questions into answers. Thanks to Bernoulli, there is; see the box overleaf.

Distributions

Rather than asking particular questions about the probabilities of different outcomes, it can be more useful to see at a glance every possible outcome. For instance, if we were keen on gambling on dice (as many people still were in the Bernoullis' time), it might be very handy to have a list of every possible score, and the probability of getting that score. For two dice, the possible totals run from 2 to 12, and have these probabilities:

Total	Probability
2	1 in 36
3	2 in 36
4	3 in 36
5	4 in 36
6	5 in 36
7	6 in 36
8	5 in 36
9	4 in 36
10	3 in 36
11	2 in 36
12	1 in 36

This complete list is called a probability distribution.

THE BINOMIAL FORMULA

Tossing a coin can only have two possible outcomes, and any probability like this is called binomial ("bi" means "two" in Greek, just think of bipeds riding bicycles wearing bifocals). For any such probability, we can use Bernoulli's binomial formula; this is similar to the formula we've used before to find combinations (see more on page 20).

$$P(x) = \frac{n!}{(n-x)! \, x!} \times p^x q^{n-x}$$

P(x) is the probability of there being exactly x "successes" (which might be heads)
n is the number of "trials" (such as coin-tosses)
p is the probability of each trial being a success (for a fair coin this would be the chance of a head, which is 0.5)
q is the probability of failure (also 0.5 for a tossed coin)

We can check that this formula works by finding the probability of one head if we toss a coin once:

$$P(1) = \frac{1!}{(1-1)! \, 1!} \times 0.5^1 0.5^{1-1} = 0.5$$

Which is what we'd expect. What do you think is the probability of getting (exactly) three heads if we toss a coin six times? Is it 0.5? Bernoulli's formula tells us it is:

$$P(3) = \frac{6!}{(6-3)! \, 3!} \times 0.5^3 0.5^{6-3} = 0.3125$$

How about the probability of getting *at least* three heads from six tosses? Is that 0.5? To find out, we calculate the probability of getting three heads (which we have just done), and add that to the probability of getting four heads:

$$P(4) = \frac{6!}{(6-4)! \, 4!} \times 0.5^4 0.5^{6-4} = 0.234375$$

plus the probability of five heads:

$$P(5) = \frac{6!}{(6-5)! \, 5!} \times 0.5^5 0.5^{6-5} = 0.09375$$

plus the probability of six heads:

$$P(6) = \frac{6!}{(6-6)! \, 6!} \times 0.5^6 0.5^{6-6} = 0.015625$$

giving a total of **0.65625**.

Plotting distributions

To be even clearer, we can also set out the ways in which each possible outcome (such as obtaining a 5 from two dice) can be obtained, as shown below. The scores of each of the two dice added together is shown to check that all possibilities have been included.

Tossing coins

If you toss two coins, the possible outcomes are HH, HT, TH or TT. Two of these outcomes have just one head, and two do not. The probability of getting one head from two throws is therefore 2 in 4, which is ½, or 0.5. Similarly, the probability of getting two heads in two throws

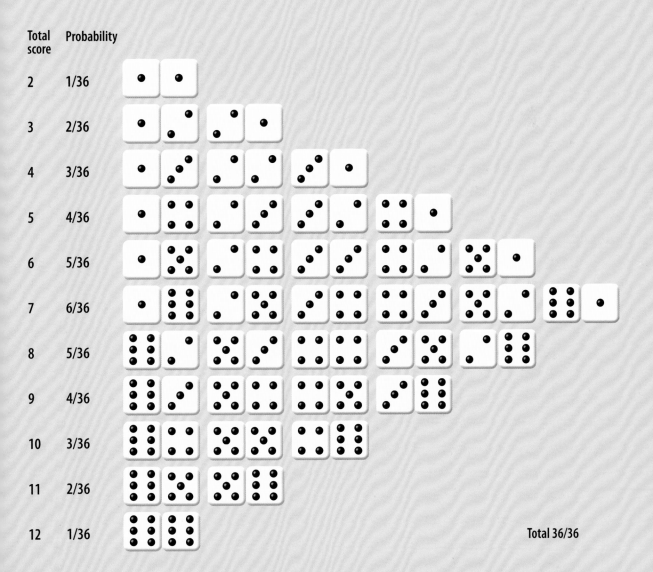

Total score	Probability
2	1/36
3	2/36
4	3/36
5	4/36
6	5/36
7	6/36
8	5/36
9	4/36
10	3/36
11	2/36
12	1/36

Total 36/36

is 0.25, and so is the probability of getting no heads. We can plot these possibilities like this:

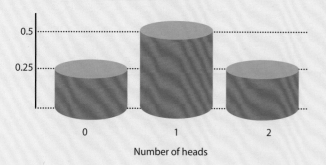

Number of heads

If we wish to explore what happens when we toss more coins, we will soon get bored of listing and counting the outcomes. Tossing one coin gives two possible outcomes (H, T), two coins gives four as above. Three coins gives eight possibilities (HHH, HHT, HTH, HTT, THH, THT, TTH, TTT), four gives 16. These numbers of outcomes are given by multiplying 2 by itself n times, where n is the number of coins.

A quicker method

So, listing the possibilities of five coins would give us 2 x 2 x 2 x 2 x 2 = 32. This can also be written as 2^5. Tossing 20 coins gives 2^{20} possible outcomes, which is 1,048,576. Fortunately, there is a quicker way—we can use a simpler form of the binomial formula (it is simpler because the probabilities of "success" (head) or "failure" (tail) are equal, so we can leave out the $p^x q^{n-x}$ part).

Probability of k heads with n coins tossed equals

$$\frac{n!}{k!(n-k)!}$$

"Probability of k heads with n coins tossed" is usually abbreviated as:

$$\binom{n}{k}$$

We can check this for the two-coin toss example.

Number of ways to get no heads with two tosses,

$$=\binom{2}{0}=\frac{2!}{0!(2-0)!}=\frac{2}{2}=1$$

Number of ways to get one head with two tosses,

$$=\binom{2}{1}=\frac{2!}{1!(2-1)!}=\frac{2}{1}=2$$

Number of ways to get two heads with two tosses,

$$=\binom{2}{2}=\frac{2!}{2!(2-2)!}=\frac{2}{2}=1$$

From scores to probabilities

To change these numbers of heads into probabilities, we just need to remember that we are certain to get either 0, 1, or 2 heads, and "we're certain," in terms of probability, means 1. So, we divide our numbers of heads by 4 to give their individual probabilities, and check that adds up to 1. We get

Zero heads one time in 4, which is 1/4 or 0.25

One head two times in 4, which is 2/4 or 0.5

Two heads one time in 4, which is 1/4 or 0.25

And these fractions add up to 1. So, now we have checked the formula works, we can use it to explore what happens when we toss more coins. If we toss 20 coins, and calculate and plot the probabilities of getting 1, 2, 3 …, 40 heads, we get Graph 1. For 200 coins the data looks like Graph 2. With an enormous number of coin tosses we would get a pattern with the outline shown in Graph 3. We get this same result whenever we are dealing with random situations in which there are just two possible outcomes, like a success or a failure, a win or lose, a head or a tail. In every case, the pattern that emerges reveals the binomial distribution.

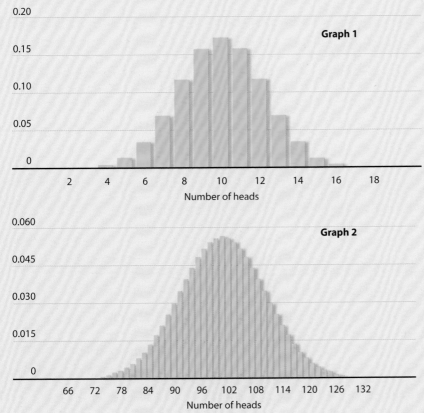

Graph 1

Number of heads

Graph 2

Number of heads

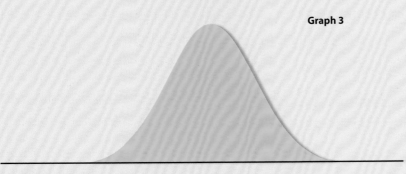

Graph 3

SEE ALSO:
▶ The Normal Distribution, page 46
▶ Seeing Statistics, page 64

The Normal Distribution

THERE ARE TWO WAYS IN WHICH DATA CAN BE COLLECTED: BY COUNTING, SUCH AS THE SCORES IN A GAME OF CHANCE, THE NUMBER OF FAULTY ITEMS in a batch, or the numbers of children in families; or by measuring, like the weights of cakes or people's blood pressures.

Items that are counted are called discrete variables; things that can be measured are continuous variables. The binomial distribution (see more, page 42) is used for discrete variables. In many cases, when continuous data is collected, a particular shape appears which is similar to, but not quite the same as, the binomial distribution. It is called the normal distribution (or the Bell curve, or the Gaussian distribution after Carl Gauss, who investigated it in great detail).

Normal cake

Let's say you have a cake-making machine, which automatically weighs each cake. You might get results (in ounces) like this:

10, 12, 11, 10, 13, 12, 14, 12

If we put the weights in order:
10, 10, 11, 12, 12, 12, 13, 14

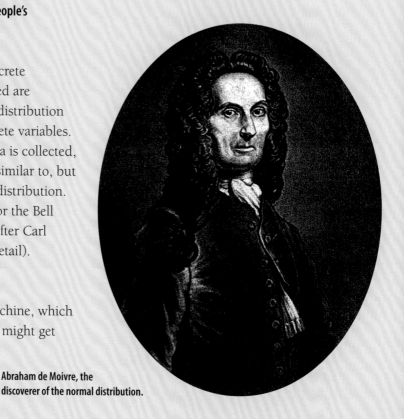

Abraham de Moivre, the discoverer of the normal distribution.

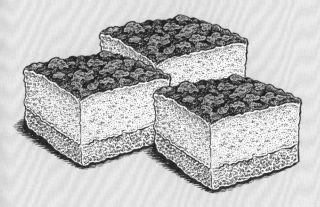

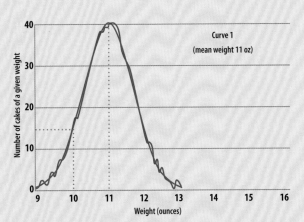

And then count the number of cakes of each weight:
10 oz : 2 cakes, 11 oz : 1 cake, 12 oz : 3 cakes, 13 oz : 1 cake, 14 oz : 1 cake

We can then plot them, drawing a chart with the weights along the horizontal axis and the number of cakes of each weight up the vertical axis. With just eight data points it won't be very informative, but if we go on measuring, sequencing, and counting until we've processed 1,000 cakes, we might get the red wiggling line called Curve 1. Shown in blue is the ideal version of such a curve, a plot of the normal distribution. To find out, say, how many of the 1,000 cakes weigh 10 ounces, you simply draw a vertical line up from the 10 entry, see where it meets the curve, and then take a horizontal line from there to the vertical axis, where you will find that there are about 15 cakes of this weight.

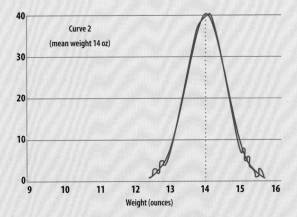

Limitations of the mean

Why might you want a plot like this? Let's say you want to sell your cakes. So, you will need to say how much they weigh. You may know that your 1,000 cakes weigh 700 pounds in total, but

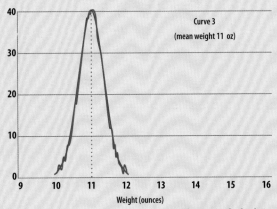

Three sets of cake data all share the same shape of curve.

it's no good just telling your customers that their average weight is 700/1,000 = 0.7 lbs = 11 oz (approximately). If someone buys one of the cakes that weighs only 9 ounces, they are not going to be happy, not even if you explain to them that that just goes to show that the average is not always all you need to know about a set of data. They may even remark "but that's just mean."

Data solutions

There are several ways around this problem. You could increase the average weight per cake to 14 oz, which would shift the distribution to the right as seen in Curve 2. You could then sell your cakes as weighing "at least 12 oz." An alternative approach is to use the chart to improve your operation, maybe by adjusting the machine to make cakes that are more similar in size. Once you have done this and repeated your weighing-and-plotting process, you will get a narrower and sharper shape, something more like Curve 3. In each case, the blue lines are those of the closest normal distribution curve to the data. The point here is that the underlying shape of the distribution of this kind of data is still a normal distribution, however it might be stretched or squeezed. (How the width of a normal distribution is defined will be covered later.)

A plot of our stock of cakes presented as a bar chart.

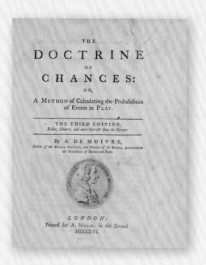

The Doctrine of Chances, a book by de Moivre first published in 1718.

Normal probabilities

The normal distribution has many uses in statistics. Looking again at the plot of the cakes below, we can think of the area under the curve as representing all those 1,000 cakes. This means we can find out that, for instance, there are 14 cakes that weigh under 9 oz. We can also think of this as a probability: there is a 14 in 1,000 chance (that is, a probability of 0.014) that a randomly selected cake weighs less than 9 oz. If we had

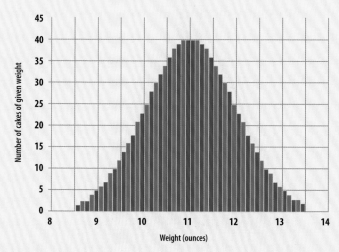

The Revocation of the Edict of Nantes (below), was a religious law passed in France in 1685. It increased the chances that de Moivre would be put to death. So he left France forever.

time to count them, we would find that there are 500 cakes weighing under 11 oz, which is to say there is a 0.5 probability that a randomly selected cake weighs less than 11 oz. Finally, since all of the cakes are plotted somewhere on the graph, which ends at 14 oz, that means that 100 percent of the cakes weigh less than 14 oz, and that the probability of a cake being somewhere in the graph is 1—a certainty.

Probabilities

What all this means is that areas under the graph are actually probabilities, and that in turn means that the normal distribution can be used to work out probabilities from those areas.

The next plot shows how this works. In any large set of data which is distributed normally, 68.2 percent of the data will lie in the blue area. So, when we apply this to our cake data (left), we can see that about 682 of the 1,000 cakes will weigh between 10 and 12 oz.

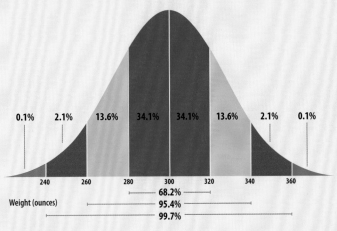

Mathematical exile

The normal distribution was discovered by Abraham de Moivre, who was born in France in 1667. He spent much of his childhood and youth learning mathematics. To begin with he taught himself by reading math books (in secret, some say), including Christiaan Huygens' book about the statistics of games of chance (see more p. 28). In 1684 de Moivre finally started to attend formal classes in mathematics and physics.

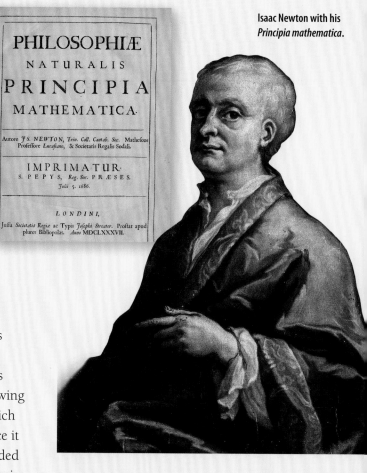

Isaac Newton with his *Principia mathematica.*

PHILOSOPHIÆ
NATURALIS
PRINCIPIA
MATHEMATICA·

Autore JS. NEWTON, Trin. Coll. Cantab. Soc. Mathefeos
Profeſſore Lucaſiano, & Societatis Regalis Sodali.

IMPRIMATUR·
S. PEPYS, Reg. Soc. PRÆSES.
Julii 5. 1686.

LONDINI,

Juſſu Societatis Regiæ ac Typis Jofephi Streater. Proſtat apud
plures Bibliopolas. Anno MDCLXXXVII.

Religious differences

De Moivre was a Protestant, and France was a Catholic country which, to put it mildly, disapproved strongly of Protestants. But this caused him few problems when he was growing up, thanks to the Edict of Nantes, a law which had guaranteed tolerance of Protestants since it had been passed in 1598. The Edict had ended the religious wars which had raged for centuries before then. But then, on October 22, 1685, King Louis XIV revoked the Edict. Hatred of Protestants meant that many were imprisoned—including de Moivre, who was only 18. When he was finally released in 1688, he fled to London, England, never to return to France.

Fun with Newton

One of the first things de Moivre did when he arrived in England was to catch up on his reading of mathematics books, and he was soon flicking through Isaac Newton's *Mathematical Principles of Natural Philosophy*. This is not an easy read, in fact one student is supposed to have muttered, when Newton passed him in the street, "There goes the man that writ a book that neither he nor anybody else understands." So, it was just the sort of light reading de Moivre loved best. He was working as a math tutor at the time, and he took to cutting out the pages of Newton's book and studying them as he traveled between students (which is definitely not recommended

THE CENTRAL LIMIT THEOREM

When data is normally distributed, the mean of the data lies in the middle, so all you need to do is to collect enough data points (30 is plenty) and average them, and you can be confident that the result you get is close to the mean of the actual population. But what if you want to find the mean value of something that is not normally distributed, such as the prices of houses, shown here.

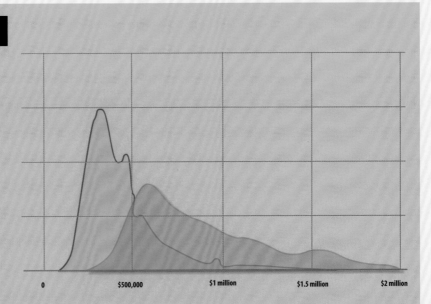

If you take a random sample in each case, will the mean be approximately the same as the true one? Thanks to de Moivre, we know that it will. He proved the central limit theorem, which shows that data randomly selected from almost any population will be distributed normally, even if the population data is not.

The blue curve shows the price of houses in the countryside. The green one shows prices in the city.

these days: a first edition of the *Principia* is worth about $3,000,000). De Moivre became such an expert on the book, and such good friends with its author, that when people pestered Newton for explanations of the math in it he would say "Ask Mr. de Moivre, he knows more about it than I do."

A big sleep

According to some biographers, de Moivre, like Cardano on page 18, successfully predicted the date of his death. Being more scientific than

Cardano, he used mathematics to do it rather than astrology. He found that he was sleeping 15 minutes longer each night and used this to calculate that he would sleep for 24 hours on 27 November, 1754. And sure enough, after that date he never woke again.

SEE ALSO:
▶ Poisson's Distribution, page 100
▶ Non-Parametric Statistics page 140

Bayes' Amazing Theorem

Reverent Thomas Bayes created the field of Bayesian probability in the 1740s.

STATISTICAL THINKING DOES NOT COME NATURALLY TO HUMAN BEINGS, and we find probabilities particularly hard to cope with. If a doctor tells you there is a 40 percent chance you have an illness, is that good news or bad? If train A has a one in a million chance of crashing, and train B has a one in a thousand chance, would you pay ten times more to travel on train A?

David Hume weighed up the likelihood of a miracle transgressing a law of nature, and the law of nature being true, in his 1748 book *An Enquiry Concerning Human Understanding*.

AN ENQUIRY

CONCERNING

HUMAN UNDERSTANDING

AND SELECTIONS FROM

A TREATISE OF HUMAN NATURE

BY

DAVID HUME

WITH HUME'S AUTOBIOGRAPHY AND
A LETTER FROM ADAM SMITH

—

CHICAGO
THE OPEN COURT PUBLISHING CO.
1921

Grappling with probabilities becomes even trickier when they are linked together, and this can be a particular problem when medical diagnoses and treatments are concerned. Often, medical tests can only show that a patient probably has a particular disease. Added to this is uncertainty about the effect of the treatment. Should a very expensive medicine be used if the probability of it working is only slightly higher than that of a cheaper one? What about using a powerful medicine with a high risk of side-effects? Or a new medicine which has only been used a few times before? The statistical formula which is used to solve problems like these is called Bayes' theorem.

Good news or bad?

Imagine there is a new disease, from which 15 million people around the world are suffering (that is 0.2 percent of the world's population of

7.5 billion), and it is turning up in every country. A test is available but it is not completely reliable. If someone has the disease, the test will be positive 99 percent of the time (and will wrongly give an all-clear 1 percent of the time; this is called a "false negative"). On the other hand, it sometimes gives a positive result when there is no disease present; these "false positives" occur in 5 percent of cases.

TRUE POSITIVE: If someone has the disease there is a 99 percent chance they will get a positive test result.

FALSE NEGATIVE: If someone has the disease there is a 1 percent chance they will get a negative test result.

FALSE POSITIVE: If someone does not have the disease there is a 5 percent chance they will get a positive test result.

TRUE NEGATIVE: If someone does not have the disease there is a 95 percent chance they will get a negative test result.

Person A has the test, and it is positive. What are her chances of having the disease?

Before the evidence

We start with the chance of Person A having the disease before we knew the test result. That is the same chance as anyone else, which is 0.2 percent. Next, we work out the chance of Person A's test result (or that of anyone else) being positive. There are two ways this can happen:

A. She has the disease and gets a true positive test result. The probability is 0.2 percent x 99 percent, which is 0.198 percent.
B. She does not have the disease, but gets a false positive. Probability is 99.8 percent x 5 percent, which is 4.99 percent.

We add the probabilities of situations A and B. The total is 5.188 percent. This is the chance of Person A receiving a positive test result.

"I say, look at that! What are the chances?"

Following a mid-air collision, this hydrogen bomb was lost at sea in 1966. It was found 80 days later, using Bayes' theorem.

After the evidence

Now, we consider the fact that Person A has had the test, and it is positive. How has this changed the probability? Now that we are dealing with probabilities that change according to the situation, we need a new symbol, which is a vertical line |. "|" means, roughly, "given that." For instance, the probability (P) that Person A has the disease (X), given that her test is positive (Y), is written **P(X|Y)**. This is what we want to work out. What we know is the opposite of this, the probability that Person A's test is positive, assuming she does have the disease. This is written **P(Y|X)**, and has the value 99 percent. We also know P(X), the probability that Person A has the disease, which is 0.2 percent. And the probability that Person A (or anyone) has a positive result, P(Y), is 5.188 percent.

The formula to work out **P(X|Y)**, is

$$P(X|Y) = P(Y|X) \times \frac{P(X)}{P(Y)}$$

Inserting the values we know:

$$P(X|Y) = 99\% \times \frac{0.2\%}{5.188\%}$$

In working calculations with probabilities, it is simpler to use fractions rather than percentages. So, we can rewrite this as:

$$P(X|Y) = 0.99 \times \frac{0.002}{0.05188} = 3.8165\%$$

So, the probability that Person A has the disease has increased from 0.2 percent before she had the test, to 3.8165 percent now that she has had a positive test result. That is to say, she still very probably does not have the disease! It is because of this that doctors rarely rely on a single test for a serious but rare disease. When probabilities change, they are called conditional. The initial probability in the formula is called the "prior," and the updated probability is the "posterior."

Useful but unused

The formula above is called Bayes' theorem and it takes its name from the Reverend Thomas Bayes, who studied something very like it in the 1740s to try to prove the existence of God and the miracles He performed. Otherwise, Bayes seems to have had little interest in the math. He was wary of probability theory because of its close associations with gambling. A few decades later, Pierre-Simon Laplace (see more, page 84) devised the formula above—and, again, did little with it. (Laplace also hated gambling.)

Real-world impact

Even when people understood that there was a lot more to probability than helping gamblers

How it works

A visual approach

The links between catching a disease and the success of tests for it can be shown like this: the area of the large blue square is the world's population, (7.5 billion) the medium yellow square is the number of people who would test positive (389 million), and the smallest square (green and red) is the number of people with the disease (15 million). The red line represents the people who have the disease but test negative, 218,000. The green area is the 14.7 million people who have tested positive and have the disease. The small size of this compared to the medium square corresponds to the low probability of Person A having the disease.

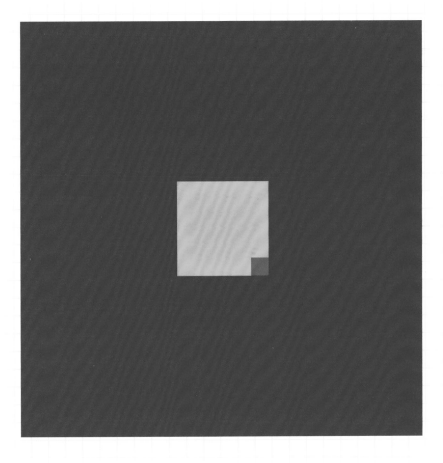

Dreaming of a white Christmas?
Bayes' theorem is your best bet
at predicting it correctly.

problem of prior probabilities. To find out the probability of a white Christmas, an obvious way to start is to check weather records and find out how frequently white Christmases have occurred in the past. You could safely use this as your prior in a Bayes' calculation (perhaps to look at the way global temperature changes affect this probability). But what about the probability of humans landing on Mars? The frequency with which this has happened in the past is zero, but if you were to put that value into Bayes' theorem, it will predict that people will never land on Mars. But clearly that isn't true. Astronauts are hoping to reach Mars this century.

What are the chances
that Martians also know
about Bayes' theorem?
What do they call it?

win, the theorem was still neglected—but gradually, more and more people realized that real-world problems very often involved conditional probabilities, and began to use the theorem. Since then, Bayes' theorem has been used to predict nuclear accidents, and to help find lost ships, planes, and nuclear warheads. Insurance companies rely on it, and so do spam filters and search engines.

The problem of the priors

Yet for many decades, the theory was widely disliked by statisticians. The reason was the

GOD. WHAT ARE THE CHANCES?

The Reverend Bayes' statistical investigations were published in 1763 as *An Essay Towards Solving a Problem in the Doctrine of Chances*, which is a reference to de Moivre's book (see more, page 46). The problem was that of reverse probability, and the best way to understand it is through the particular example that interested Bayes. Bayes wanted to use the existence of the Universe to prove the existence of God. (This argument has a long history in philosophy and theology and is known as the cosmological argument.) If this could be proved, then it could be reworded as a statement of probability: "Given that the Universe exists, the probability of the existence of God is 100 percent." In symbols, this would be $P(G|U) = 1$. Bayes' idea was to begin with what is called the "reverse probability," and this is closely linked to the conditional probabilities of Bayes' theorem. Since God (at least, the Christian version that Bayes was interested in) is defined as the creator of the Universe, the probability that the Universe exists if God exists is 100 percent. That is $P(U|G) = 1$. The question is, can one prove $P(G|U) = 1$ from $P(U|G) = 1$? The answer is no.

No problems

People who like Bayes' theorem are relaxed about this. They say that there is more to probability than frequency and that instead, a plausible probability can be used as a starting point. A NASA engineer could easily suggest a plausible probability for a Mars landing, for example. Since the whole point of Bayes' theorem is to refine probabilities, even a dubious starting point is acceptable. Those who dislike Bayes' theorem (they are called frequentists) would say that plausibility is not good enough. In fact, the reason we need statistics is because we are so bad at estimating probabilities ourselves. It's not plausible that you will ever win the lottery, but that doesn't stop you playing it (or does it?). After decades of research, however, it became clear that Bayes' theorem is very useful and solves problems that cannot be tackled in any other way.

SEE ALSO:
▶ What to Expect, page 28
▶ Correlation, page 122

The Needle of Chance

MATHEMATICS IS FULL OF UNEXPECTED CONNECTIONS, and one of the most surprising is between chance and the geometry of circles.

Imagine a table decorated with lines spaced out an inch apart. Drop a 1-inch needle on the table and it will either land so that a part of it crosses (or touches) a line, or not. Keep dropping the needle and note the number of times it crosses or

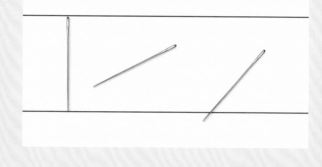

Georges-Louis Leclerc, Comte de Buffon, discovered the link between π and probability.

touches a line. Call this number C, and call the total number of drops N. Keep going until N = 1,000, and calculate a value for 2N ÷ C. You will find that the answer is (more or less) the ratio of the circumference of a circle to its diameter, which is 3.14159…, symbolized by the Greek letter π (pronounced "pi"). If you think dropping a needle a thousand times is more than can be expected of even a dedicated mathematician, bear in mind that Georges-Louis Leclerc, Comte de Buffon, who discovered this strange way of estimating π, actually tossed a coin 4,004 times, just to check that it did indeed come up heads about half the time (he got 2,028 heads—and 1,976 tails).

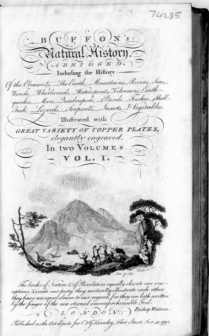

Nature lamenting over the Tomb of M. De Buffon, &
exhibiting a Portrait of that distinguished Naturalist.

An English translation of Buffon's *Natural History*, published in 36 sections between 1749 and 1804.

On the run

Buffon began to study mathematics in 1728, when he was 21, but had to leave without completing his course following a duel. The details are mysterious but he seems to have fled, first to various places in France and then on to Italy. He stayed there until his mother died and he inherited the family wealth. At this point, he returned home with so much money that he could do whatever he liked. (And what he liked was mathematical science, so he spent the rest of his life studying it.)

A world without a God

Very unusually for the time, Buffon also attempted to explain the origin of the Universe and of life through purely scientific principles, without any place for God. He was one of the first to argue that life formed from chemicals, that humans are animals, that animals have thoughts and feelings, and that the Solar System was formed through natural processes. He was right every time, but the data and scientific theories to prove it would not be discovered for many decades after his death. In 1749, Buffon set out to write his account of life, the Universe, and just about everything else. Realizing that there was no time to waste, he decided he ought to get up early—five o'clock in the summer, six o'clock in winter. He instructed his valet, Joseph, to make sure he did this, and paid him extra if he did. Unfortunately Buffon was a deep sleeper, so Joseph sometimes had to drag him out of bed onto the floor. When even that didn't work, Joseph raised his master with a bucket of iced water.

MAKING EQUATIONS

In math, we can always replace the proportionality symbol ∝ with an equals sign by introducing a constant; for instance, the longer the time you walk, the further you get. In other words, distance walked is proportional to time spent walking. In symbols, this is d∝t. Now, we introduce a suitable constant. In this case, the constant is your walking speed (call it s). So, d=s × t. The benefit of turning a proportionality into an equation is that equations can answer questions like "How far can you walk in 2 hours?" If your walking speed is 3 miles per hour, the equation tells you that the distance you can walk is 3 miles per hour × 2 hours = 6 miles.

Fire from the Sun

It was not only religious myths that Buffon investigated: he was also unconvinced by the legend that Archimedes used reflected sunlight to set fire to wooden ships that were attacking his homeland (the city of Syracuse on what is now the island of Sicily), in 212 BCE. So, Buffon bought 168 mirrors and used them to reflect the Sun to form a heat ray, with which he successfully set fire to piles of wood at a distance of 150 feet!

The noodle of chance

The simplest way to show why π turns up in Buffon's needle experiment is to solve what

Archimedes is said to have set the invaders' fleet ablaze with focused sunlight.

sounds like a trickier problem. Imagine dropping a noodle (a cooked and flexible one) on your ruled paper, and let's say the noodle can be of any length. We are going to work out the number of times that we expect the noodle to touch a line. We write this expectation as E(A). If the noodle is several feet long, there will be lots of points where it crosses a line, so E(A) must be quite big. If the noodle is very short indeed, it's very likely to fall into a gap between lines, which means E(A) must then be about zero. In other words, the longer the noodle, the larger E(A) becomes. We can write this as

E(A) for a noodle of length l ∝ l

(∝ means "proportional to").

Or, more briefly, as **E(l)∝l**

To change our noodle expectation expression **E(l) ∝ l** into an equation, we need to insert a constant; **E(l) = k × l**. We can find the value of k by turning the noodle into a circular hoop, with a diameter (d) which is the same as the gap between the lines.

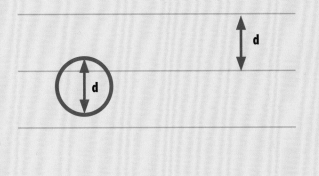

The hoop will almost always cross a line in two places. That is **E(hoop length) = 2**. The length of the hoop is given by the equation for the circumference of a circle, **c = π × d**, so **E(hoop length) = 2** becomes **E(π × d) = 2**. Rearranging this gives us our constant **k = 2/πd**, and putting this in our previous expression changes it from **E(l)∝l** to **E(l)=2/πd × l**. And that's where the π value appears from.

The Monte Carlo Casino attracts gamblers and statisticians.

Monte Carlo or bang

Buffon's needle experiment is an early example of a technique now called a Monte Carlo method. It is called this because it's a kind of game of repeated chances, like roulette and other games famously played at casinos in Monte Carlo, the main city in the tiny European country of Monaco. The name was coined, and the approach clearly defined, in 1946 by a Polish

mathematician called Stanislaw Ulam, who at the time was working on the secret U.S. project to develop a nuclear reactor.

The neutron problem

Nuclear reactions all depend on the way in which tiny particles called neutrons interact with atoms. Neutrons are released by radioactive materials, and, once released, sometimes knock other neutrons out of other atoms. If these further neutrons release even more, a steady flow of neutrons results, and a lot of heat is generated by nuclear fission. This heat can be used to make electricity, and the whole system is a nuclear reactor. However, if each neutron knocks out

Stanislaw Ulam shows his daughter the controls to the MANIAC (Mathematical Analyzer Numerator Integrator And Computer).

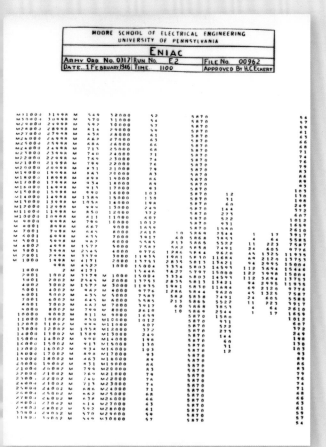

ENIAC (left) was programmed and fed with data by punched cards and gave out its results in the same way. The results were then printed into data as shown here. In Ulam's Monte Carlo simulations, each neutron was represented by a single punched card.

to one of the world's first electronic computers, the ENIAC (Electronic Numerical Integrator And Computer) and, soon after, to upgraded devices called FERMIAC and MANIAC, the math was still too difficult.

Work through play

Ulam liked to play solitaire, and one evening he started to ponder how he might work out his chances of winning a game. Like his neutron problem, he could do it if he knew enough about the mixture of playing cards he was using, and had time to work out every possible way of winning and losing. But, then he realized that another way would be just to play many games. This might take longer than the calculations, but would be much simpler. Ulam knew that this kind of simple, but time consuming, approach was just what computers were good at: while the ENIAC could not handle the complicated neutron calculations, it could easily predict the random motion of a single neutron. By making these predictions over and over again, and seeing which led to nuclear explosions, it then became easy to estimate the probability of an explosion given a particular reactor design. Monte Carlo methods are still very popular with nuclear physicists today.

more than one new neutron, a runaway process begins which very soon turns a nuclear reactor into a nuclear bomb! Working out whether this will happen depends both on knowing a lot about the mixtures of materials present, and on the answers to some very tricky calculations. Even though Ulam and his colleagues had access

SEE ALSO:
▶ Odds, page 14
▶ Tests and Trials, page 160

Seeing Statistics

IT SEEMS SO NATURAL TODAY TO DISPLAY STATISTICS IN THE FORM OF CHARTS OF VARIOUS KINDS, that it's hard to imagine doing without them. Yet, until the 1820s, such visual aids hardly existed.

Three of the most useful charts today are the bar chart, the line chart, and the pie chart, and all of them were invented by the same person, William Playfair. In fact, the world's first bar chart (see top right) was also the world's first line chart. As is the case in many modern charts, time proceeded along the horizontal axis from left to right. The labels on the left vertical axis, and the vertical bars, show the price of wheat in London, and a red line and the right vertical axis show workers' wages.

A muddle of meanings

Like many charts, this one was intended to explain a particular point to its viewers, but unfortunately, this first example was a complete failure.

William Playfair and his trade-balance time-series chart, published in his *Commercial and Political Atlas* of 1786.

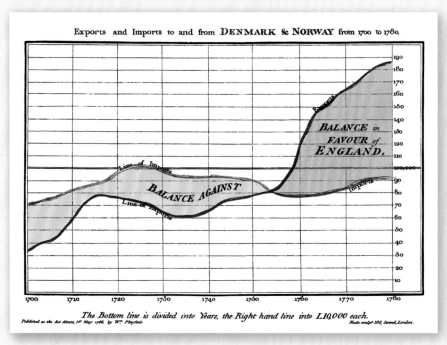

Exports and Imports to and from DENMARK & NORWAY from 1700 to 1780.

BALANCE in FAVOUR of ENGLAND.

Line of Imports

BALANCE AGAINST

Line of Exports

Imports

The Bottom line is divided into Years, the Right hand line into £10,000 each.

Published as the Act directs, 14 May 1786, by Wᵐ Playfair

Neele sculpt 352 Strand, London.

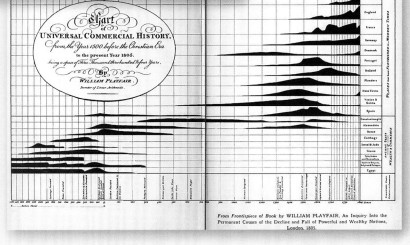

Left: Playfair's pioneering chart or "Chart Showing at One View the Price of the Quarter of Wheat, and Wages of Labour by the Week, from 1565 to 1821."

Below: "Chart of Universal Commercial History" from 1805.

From Frontispiece of Book by WILLIAM PLAYFAIR, An Inquiry Into the Permanent Causes of the Decline and Fall of Powerful and Wealthy Nations, London, 1805.

What it is supposed to show is that, while wheat increased in price over the period, workers' wages increased faster and so they were becoming gradually better off (at least in terms of the wheat they could afford). But this is not really clear. A much better way to show this is by plotting a different statistic, such as the number of weeks required to buy a large bag of wheat.

Perfect plotting

Some of Playfair's charts could not be bettered, however. The two-line graph on the left plots the total UK imports from Denmark and Norway against time as well as total UK exports to those countries. It shows very clearly how the difference between the two (the trade balance) has changed from negative to positive over the period, and could hardly be clearer or simpler, including only the information that is relevant and using color very effectively, too.

A new kind of plot

At the time, few people adopted Playfair's new displays, despite the fact that he published them in books specially designed to make them attractive, and even colored them all in himself by

A VERY SMALL PIE

Playfair is credited with drawing the first ever pie chart, though it was so tucked away on one of his many diagrams that he must not have been especially excited by it. He used it to show how much land the Ottoman Empire of Turkey had once occupied in Africa, Asia, and Europe.

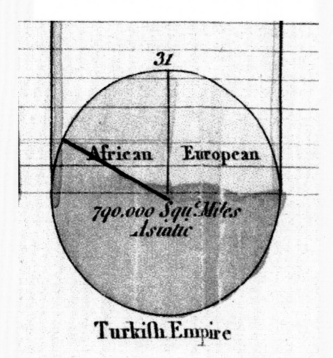

of inventor and national hero James Watt, Playfair was apparently both a blackmailer and a spy, and in 1793 he even tried (with some success) to wreck the French economy by flooding its markets with counterfeit bank notes.

The facts made clear

So, it was very fortunate for the history of statistical charts that one of the next people to come up with a powerful new graphical way of presenting statistics was Florence Nightingale, one of the most beloved and well-known women of her day (or of any other), thanks to her pioneering work as a nurse in the Crimea during the 1853–56 war between France and Britain and their allies on one side, and Russia and its empire on the other. When Nightingale discovered that more soldiers were dying from infections (caused partly by the insanitary conditions in the field hospitals of the day) than in battles, she was determined to make politicians and military leaders take the problem seriously, and she used a new kind of statistical chart, based on the pie chart. Nightingale went on to become the first woman to be elected as a Fellow of the Statistical Society of London (in 1858), and she also argued for the idea of a medal for achievement in statistics, in memory of Adolphe Quetelet (see more, page 96).

hand. The problem was that, although for once statistics was being presented in way that had nothing to do with games of chance or anything else disreputable, William Playfair had a very bad reputation himself. This seems to have been thoroughly well-deserved. Despite being a friend

SEE ALSO:
▶ The Shapes of Data, page 38
▶ Measuring Confidence, page 154

DIAGRAM of the CAUSES of MORTALITY
IN THE ARMY IN THE EAST.

2.
APRIL 1855 to MARCH 1856.

1.
APRIL 1854 to MARCH 1855.

The Areas of the blue, red, & black wedges are each measured from the centre as the common vertex.

The blue wedges measured from the centre of the circle represent area for area the deaths from Preventible or Mitigable Zymotic diseases, the red wedges measured from the centre the deaths from wounds, & the black wedges measured from the centre the deaths from all other causes.

The black line across the red triangle in Nov.ʳ 1854 marks the boundary of the deaths from all other causes during the month.

In October 1854, & April 1855, the black area coincides with the red; in January & February 1855, the blue coincides with the black.

The entire areas may be compared by following the blue, the red & the black lines enclosing them.

Above: In Nightingale's graph, each segment represents one month. The distances from the center to the edges of colored sections show the number of deaths: pink is battle, blue is infection, and black is "other causes." So, even in the most violent month (November 1854), the number of deaths in battle is only about half of that due to infection.

Left: A military hospital ward during the Crimean War, prior to the interventions of Florence Nightingale.

Measures of Spread

THE NORMAL DISTRIBUTION GRAPH SHOWS THE WAY IN WHICH MANY KINDS OF DATA TEND TO ARRANGE THEMSELVES (see more, page 46). The normal distribution curve turns with so many kinds of data, it is useful to describe its width— or spread—in a general way.

Carl Friedrich Gauss developed standard deviation in 1821.

If we could describe the width of a normal distribution graph in a general way, one that we could use with any kind of data, we could then compare different normal distributions. To do this, we can introduce a new unit, "standard deviation," symbolized by a sigma (σ), the Greek letter s. Adding sigma labels to a normal distribution creates a graph divided up as shown on the top of the opposite page.

Sigma in science

Standard deviation is the most popular way of describing how spread-out data is. Together with the mean, the standard deviation is all that is needed to define a normal distribution. Standard deviation is often used in science. When looking for evidence of new kinds of subatomic particles, for instance, particle physicists have to cope with the problem that all sorts of random clicks and flashes can appear in their equipment. So, they cannot be certain that what they hope is a new particle is not actually an unusually bright, random flash. By considering all kinds of other data, they can draw a conclusion like "there is a 1 in 99.73 chance that this is a new particle." As the top right diagram shows, this corresponds to 3σ, which is often simply referred to as a "three sigma chance." In fact, because they want to be very confident that they are

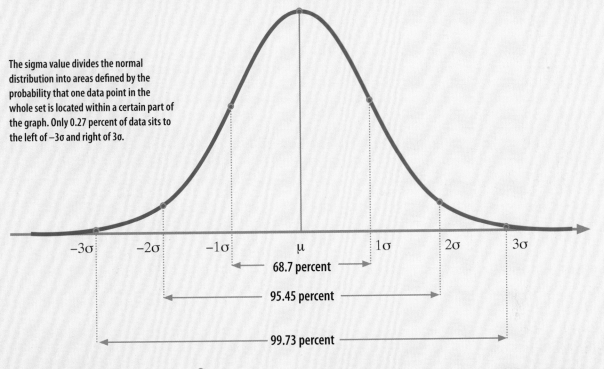

The sigma value divides the normal distribution into areas defined by the probability that one data point in the whole set is located within a certain part of the graph. Only 0.27 percent of data sits to the left of −3σ and right of 3σ.

-3σ -2σ -1σ μ 1σ 2σ 3σ

68.7 percent

95.45 percent

99.73 percent

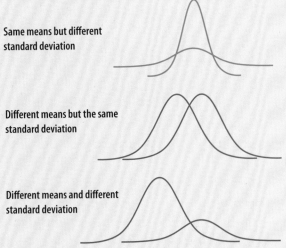

Same means but different standard deviation

Different means but the same standard deviation

Different means and different standard deviation

Left: Normal distributions can differ from one another only in terms of their means and their standard deviations.

Below: The Large Hadron Collider's advanced detectors must have their results confirmed using math developed in the 1820s.

correct, they will usually only believe and announce a new discovery that is even more unlikely to be due to chance. When the discovery of the weird new particle, the Higgs boson, was announced in 2010, it was supported by observations at about the 5σ level. This corresponds to a probability of about 0.00003 percent (about 1 in 3 million). That is to say, the chance of a random flash causing the observation, rather than an actual particle, was about 1 in 3 million.

Gaussian curves

The standard deviation was first used by German mathematician Carl Friedrich Gauss in 1821, though it was not given its name until the 1890s (Gauss called it the "mean error"). Gauss is one of the greatest of all mathematicians, and was very well aware of that fact. He even forbade his sons to

become mathematicians, because of the risk that their poor performance compared to his own might be bad for his reputation; as he put it, it might "sully (or dirty) the family name."

More at home with the range?

Just as the mean is only one of several ways to find the middle of a set of data (others are the mode and median; see more, page 10), so standard deviation

Having been banned from math by their high-achieving father, Wilhelm Gauss (above) became a shoe seller in St Louis, Missouri, Eugene (right) founded a bank in St Charles, Missouri, while Joseph (unpictured) was a soldier and railway engineer in Germany.

THE SECOND SIGMA, AND OTHER SPECIAL SYMBOLS

Rather than writing "the mean of the whole population," statisticians often use the Greek letter **μ** (pronounced "mew"), and, instead of "the mean of the sample data," they write **x̄**. Another useful symbol is the Greek letter **Σ** (this is the upper case version of **σ**, so is also called "sigma"). It means "sum the following," and we can use it to write an equation for the standard deviation:

$$\sigma = \sqrt{\frac{\sum_{i=1}^{N}(x_i - \bar{x})^2}{N}}$$

Which means, in words, "for a set of data with **N** members, take each value in turn (symbolized by $x_1, x_2, x_3, x_4 \ldots x_N$) and subtract the mean from that value. Square each answer. Add all the answers together, divide by **N**, and finally work out the square root of the whole thing."

How it works

The sigma of snow

To see where the standard deviation comes from and how it is calculated, let's look at some real data. To keep the calculations manageable, we will look at just four pieces of data, even though standard deviation is usually calculated for much larger samples. Mount Washington in New Hampshire is the snowiest place in the USA. In the years 2000 to 2003, it had the following numbers of snowy days: 108, 106, 128, and 130. To find the standard deviation we first work out the mean.

Meteorologists have been collecting snow data at the Mount Washington Observatory since 1932.

$$\frac{(108 + 106 + 128 + 130)}{4} = 118$$

How might we define the spread of the data? One obvious way to do this is to work out the difference of each number from the mean. These differences are

2000	$118 - 108 = 10$
2001	$118 - 106 = 12$
2002	$118 - 128 = -10$
2003	$118 - 130 = -12$

However, if we average these values, we find

$$\frac{(10 + 12 + -10 + -12)}{4} = \frac{0}{4} = 0$$

In fact, whatever numbers we use, we will find that their average difference from the mean is zero. To avoid this we make all the values positive by squaring them before they are averaged

$$\frac{(10^2 + 12^2 + (-10)^2 + (-12)^2)}{4} = \frac{(100 + 144 + 100 + 144)}{4} = \frac{488}{4} = 122$$

122 is called the variance, and is sometimes used directly as a measure of spread. More often, statisticians prefer a number more similar to the differences, which is obtained by finding the square root. For our data, it is $\sqrt{122} = 11$ approximately. This is the standard deviation. Although the standard deviation is very often used with normal data, there is no problem with using it for other data types (in the example of the snowy days for instance, we have no idea how the data is distributed). But it is only in the case of normally-distributed data that we can relate standard deviation values to probabilities.

Delivering food whatever the weather is more risky than you might think.

is not the only measure of spread—in addition to the variance, mentioned previously, there is also the range, which is simply the difference between the largest and smallest value. For the snowy days data (see box, page 71), the range is 130 – 108 = 22. The question that is used to decide whether to use the standard deviation or the range is: are the end-values of the set of data more important / interesting than the rest? If you are interested in the day-by-day spread in temperature of the water from a shower, the hottest and coldest values are especially important to avoid scalding or freezing the person being showered. But, if you are a supermarket manager and you want to find out whether your fresh fruit suppliers will be able to offer grapes that are all roughly the same size, you won't be interested in the occasionally enormous or teeny grape. What you want to know is how much variation there is among all the grapes on offer. So, in this case, standard deviation will be much more useful than range.

Bessel's correction

Very often in statistics we don't have all the data

Friedrich Wilhelm Bessel.

there is about the thing we are interested in. A shoemaker might want to know what fraction of the world's population has size 9 feet, but she's not going to be able to find out. So instead she will have to make do with a sample; the shoe sizes of 10,000 people, for example, would probably be enough. Unfortunately, there can be a problem when using the standard deviation on sampled data, especially when the sample is much smaller than the population. Let's say that the actual mean shoe size of the world's population is 8.4, and that our small sample of shoe sizes is 8, 8, 8.5, 9.5 and 10. The correct value of the standard deviation from the mean is

$$\sigma = \sqrt{\frac{((8\text{-}8.4)^2+(8\text{-}8.4)^2+(8.5\text{-}8.4)^2+(9.5\text{-}8.4)^2+(10\text{-}8.4)^2)}{5}} \approx 0.906$$

However, what if we don't know the actual mean? The best we can do is work it out from the data we have, which gives

$$\frac{(8+8+8.5+9.5+10)}{5} = 8.8$$

Unfortunately, if we put this value in our equation for the standard deviation, we get

$$\sigma = \sqrt{\frac{((8.8\text{-}8)^2+(8.8\text{-}8)^2+(8.8\text{-}8.5)^2+(8.8\text{-}9.5)^2+(8.8\text{-}10)^2)}{5}} \approx 0.812$$

In fact, whenever we are forced to use the mean of the sample data rather than the actual mean of the population, we will always get a smaller value of the standard deviation than the correct one. A correction for this underestimate was suggested by mathematician and astronomer Friedrich Wilhelm Bessel (see pages 90–5) in 1861: instead of dividing the sum of the squared differences of our sample of 5 numbers by 5, we divide it by 4. Then we get

$$\sigma = \sqrt{\frac{((8.8\text{-}8)^2+(8.8\text{-}8)^2+(8.8\text{-}8.5)^2+(8.8\text{-}9.5)^2+(8.8\text{-}10)^2)}{4}} \approx 0.908$$

The correction doesn't always work as well as this, but it usually does give an answer closer to the correct one. If we had 6 items, we would divide by 5, and, in general, if we had n items, we would divide by n-1.

SEE ALSO:
▶ Skew, page 132
▶ Compare and Contrast, page 146

Least Squares

WHAT DO THESE PAIRS HAVE IN COMMON? Book weight and number of pages; city area and city population; engine capacity and power. One answer is that you might expect them all to be roughly proportional relationships. That is, if you double the number of pages in a book, the population of a city, or the capacity of an engine, you could expect the weight of the book, the area of the city, or the power of the engine to double, more or less.

If you've read page 61, you'll know we can change a proportional relationship into an equation by introducing a constant, so we might go from **Book weight ∝ number of pages** to **Book weight = k × number of pages.** In this case, the

Pages from *New Methods for Determining the Orbits of Comets* by Adrien-Marie Legendre, published (belatedly) in 1806.

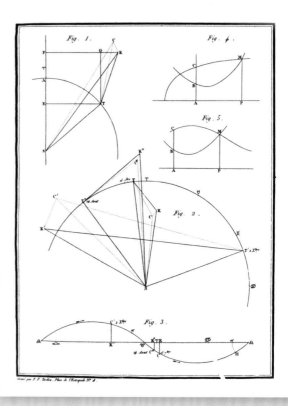

Above: Cities can get very crowded.
Left: Not all books are as weighty as these ones.

constant **s** would be the weight of a single page; if this is 2 grams, we would have **book weight=2 × number of pages**, and this could be useful to a publisher or bookbinder as it could tell them the weight of a newly planned book.

Rough solutions

None of these are likely to be exact relationships: for example, book covers weigh more than pages, people moving into a city may well move into smaller flats than established residents, and engine power depends on lots of things in addition to capacity. These facts mean that our equations are not completely accurate.

So, in these cases, we would like a way to establish the relationships, and in particular the constants, even though we will have to proceed in a rough and approximate way. This is exactly the kind of problem that statistics was designed to solve, and the simplest solution is to measure some actual books, cities, or engines, and draw a chart of the results, as seen on the right.

Defining the best

Next, we draw a straight line that best fits these points. We could do this by eye, but the data are so scattered that there is a lot of choice, so we would like a mathematical way to define what "best fit straight line" means. A simple and very effective way to do this is to say that the line with the best fit is the one which minimizes the sum of

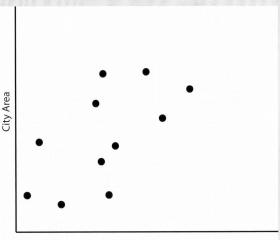

the distances of the points from it. Like this one:

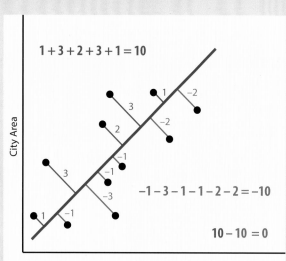

In this case, we label the distances from the line to the points that fall below the line (the green distances) as negative and the distances from the line to the points above the line (in red) as positive. This allows us to add them all up to get zero—that is the best possible fit of a straight line to this data.

Better lines

This method of finding best-fit straight lines is the simplest and clearest, but not the most popular. A problem is revealed by certain patterns of data, like this one:

This can be fitted equally well by these three lines, as the sums of their distances show:

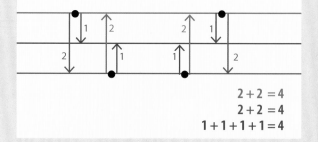

KINDS OF DISTANCE

Sometimes, rather than measuring the minimum distances of the points from the line, vertical distances are used instead, partly because they are simpler to calculate (as shown below). In either case, the distances are sometimes known as offsets. Meanwhile, minimum distances are known as perpendicular offsets, because each of them makes a right angle with the line (see these in action on page 80).

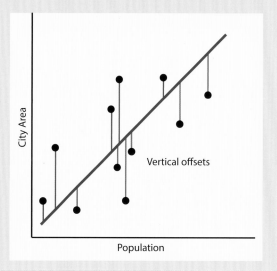

The only picture of Adrien-Marie Legendre is this caricature above. His upset espression may be due to living through the Reign of Terror, a time when torture and execution were common across France.

Least squares

There is a way around this however, which is to minimize, not the sum of the distances of the points from the line, but the sum of their squares. If we apply this approach to the last example, we can see that the blue line is now "best:"

$$2^2 + 2^2 = 8$$
$$2^2 + 2^2 = 8$$
$$1^2 + 1^2 + 1^2 + 1^2 = 4$$

Dangerous days

The least squares approach was first studied by Adrien-Marie Legendre, who had the misfortune to live in Paris during the French Revolution; in 1793 he lost his family wealth and his position in the French Academy of Sciences, and nearly had

to go into hiding for a while. But his reputation as one of the greatest mathematicians of his time meant that he was welcomed to the new version of the Academy when it reopened in 1795 as the Institut Nationale des Sciences et des Arts.

Much of what we know of Adrien-Marie Legendre comes from his colleague Siméon-Denis Poisson, pictured here.

CELEBRATING CERES

It was probably Gauss' famous scientific report on his re-location of Ceres that Sir Arthur Conan Doyle had in mind when he wrote a dramatic Sherlock Holmes story. In this detective story, Conan Doyle mentioned a science classic written by the evil but brilliant Professor Moriaty: "Is he not the celebrated author of *The Dynamics of an Asteroid*, a book which ascends to such rarefied heights of pure mathematics that it is said that there was no man in the scientific press capable of criticizing it?"

Sherlock Holmes and Professor Moriarty meet their ends in a fight over a waterfall—or do they?

THE DEATH OF SHERLOCK HOLMES.

Beobachtungen des zu Palermo d. 1 Jan.

1801	Mittlere Sonnen-Zeit			Gerade Aufstig. in Zeit			GeradeAuf Steigung in Graden			Nör Abwo
	St.	,	"	St.	.	"	.	"		. '
Jan. 1	8	43	17,8	3	27	11,25	51 47	48,8		15 37
2	8	39	4,6	3	26	53,85	51 43	27,8		15 41
3	8	34	53,3	3	26	38,4	51 39	36,0		15 44
4	8	30	42,1	3	26	23,16	51 35	47,3		15 47
10	8	6	15,8	3	25	32,1	51 23	1,5		16 10
11	8	2	17,5	3	25	29,73	51 22	26,0		. . . .
13	7	54	26,2	3	25	30,30	51 22	34,5		16 22
14	7	50	31,7	3	25	31,72	51 22	55,8		16 27
17	.	. .	.	.	. .	.	. . .			16 40
18	7	35	11,3	3	25	55,	51 28	45,0		. . . .
19	7	31	28,5	3	26	8,15	51 32	2,3		16 49
21	7	24	2,7	3	26	34,27	51 38	34,1		16 58
22	7	20	21,7	3	26	49,42	51 42	21,3		17 3
23	7	16	43,5	3	27	6,90	51 46	43,5		17 8
28	6	58	51,3	3	28	54,55	52 13	38,3		17 32
30	6	51	52,9	3	29	48,14	52 27	2,1		17 43
31	6	48	26,4	3	30	17,25	52 34	18,8		17 48
Febr. 1	6	44	59,9	3	30	47,2	52 41	48,0		17 53
2	6	41	35,8	3	31	19,06	52 49	45,2		17 58
5	6	31	31,5	3	33	2,70	53 15	40,5		18 15
8	6	21	39,2	3	34	58,50	53 44	37,5		18 31
11	6	11	58,2	3	37	6,54	54 16	38,1		18 47

Legendre greatly disliked biographers and, according to his colleague Siméon-Denis Poisson (see more, page 100) "often expressed the desire that, in speaking of him, it would only be the matter of his works, which are, in fact, his entire life." So it's not surprising that many details of his life are unknown.

The lost world

However, Legendre seems not to have found the least squares approach very interesting, and it was Carl Gauss (see more, page 70) who showed its true power. Gauss used the least squares method to track down a missing planet.

s Prof. Piazzi neu entdeckten Gestirns.

eocentri- he Länge	Geocentr. Breite	Ort der Sonne + 20" Aberration	Logar. d. Distanz ☉ ☿
3 22 58,3	3 6 42,1	9 11 1 30,9	9,9926156
3 19 44,3	3 2 24,9	9 12 2 28,6	9,9926317
3 16 58,6	2 58 9,9	9 13 3 26,6	9,9926324
3 14 15,5	2 53 55,6	9 14 4 14,9	9,9926418
3 7 59,1	2 29 0,6	9 20 10 17,5	9,9927641
. . . .	. . . .	. . . .	. . . .
3 10 27,6	2 16 59,7	9 23 12 13,8	9,9928490
3 12 1,2	2 12 56,7	9 24 14 13,5	9,9928809
. . . .	. . . .	. . . .	. . . .
3 25 59,2	1 53 38,2	9 29 19 53,8	9,9930607
3 34 21,3	1 46 6,0	10 1 20 40,3	9,9931434
3 39 1,8	1 42 28,1	10 2 21 32,0	9,9931886
3 44 15,7	1 38 52,1	10 3 22 22,7	9,9932348
4 15 15,7	1 21 6,9	10 8 26 20,1	9,9935061
4 30 9,0	1 14 16,0	10 10 27 46,2	9,9936332
4 38 7,3	1 10 54,6	10 11 28 28,5	9,9937007
4 46 19,3	1 7 30,9	10 12 29 9,6	9,9937703
4 54 57,9	1 4 1,5	10 13 29 49,9	9,9938423
5 22 43,4	0 54 23,9	10 16 31 45,5	9,9940751
5 53 29,5	0 45 5,0	10 19 33 33,3	9,9943276
6 26 40,0	0 36 2,9	10 22 35 11,4	9,9945823

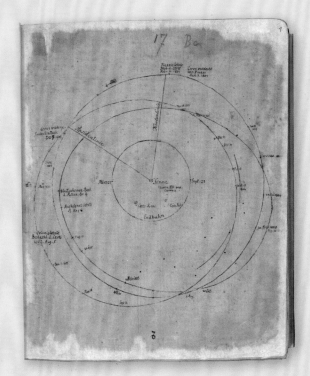

Left: The original data of Ceres' motions collected by Piazzi.
Above: Gauss's original sketch.

Below: Giuseppe Piazzi's account of the discovery of Ceres, the largest body in the Asteroid Belt.

Ceres, the fifth planet from the Sun, was discovered on the first day of 1801 by an Italian astronomer called Giuseppe Piazzi, but its orbit soon took it behind the Sun. By then, Piazzi had charted its position on 19 nights, but at the time telescopes were still not good enough to allow such positions to be measured accurately. The rough data Piazzi had were not good enough to work out the path of Ceres in the sky, and so no one knew where to look for the new planet once it had emerged from behind the Sun again. Gauss applied his least squares method to just three of Piazzi's observations, and predicted the position that Ceres would occupy once it was

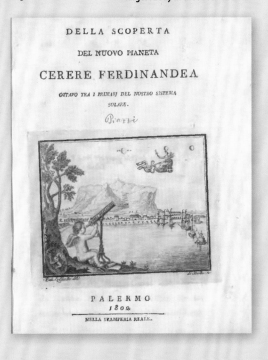

visible again. Two astronomers found the missing world at almost exactly that location, on the last day of 1801.

The fate of Ceres

If you Google "fifth planet from Sun" now, you won't find Ceres: it was reclassified as an asteroid the very next year after it was discovered. However, in 2006 it partially regained its planetary status when it was reclassified once more, this time as a dwarf planet.

Data from lines

Assuming that you are convinced that the line you have found actually means something—that is, that there is a real relationship between the data you have plotted—you can use it to find out new information. You could, for instance, find the population of a city of 1,000 square miles, or you could find the area of a city of 10 million inhabitants. To do this, you can simply draw a line from the data-point you have (say, a population of 10 million), extend that line to the best-fit line, and then draw a new line from that intersection point to the other axis—where you will find 2,000 square miles (these are the pink lines above right).

Scattering

However, in this example it would not be wise to believe that your conclusions are very reliable—judging by the distances of the points from the line (this is known as the "scatter" of the data), this is a very approximate relationship indeed, and so should your conclusions be.

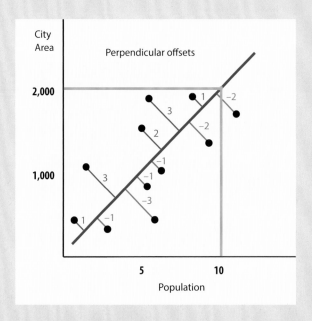

Lines without drawings

Drawing lines and reading off answers is a time-consuming process, and a tricky thing to teach computers to do. Instead, we can find the equation that describes the line: mathematically, the line and its equation are really the same thing. This is how it's done:

Step 1: Select points

First, we find two points on our straight line as seen above right. We can write these points as bracketed pairs of numbers called coordinates. The coordinates corresponding to the point where the first pair of lines meet are (10 million people, 2,000 square miles), and those of the second pair of lines intersect at the point (5 million people, 1,000 square miles).

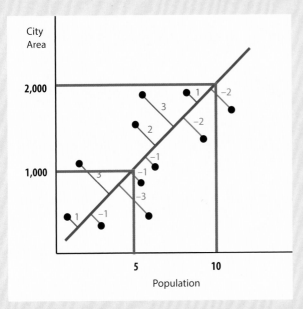

Step 2: Select a general formula

Now we write down the general formula for any straight line:

$y = mx + c$

There is a convention which says that points along a horizontal axis are called x-coordinates, and those up the vertical axis are y-coordinates, and these are what the **x** and **y** in the equation refer to. The **c** is the point where the line crosses the y-axis (which is zero in this case, as we would expect. A city of zero area can contain no people). The **m** is the slope (or gradient) of the line; **m** is zero for a horizontal line and infinitely large for a vertical line.

Step 3: Find the slope or gradient, m

We replace the x and y in the equation with the values from our points (10, 2000) and (5, 1000), giving us two versions of the equation

$2000 = m \times 10 + c$
$1000 = m \times 5 + c$
Subtract these equations
$2000 - 1000 = m \times 10 + c - m \times 5 - c$
Simplify
$1000 = m \times 10 - m \times 5 = m \times 5$
And use this to work out the value of m:
$m = \dfrac{1000}{5} = 200$

And put this m back into the equation,
$y = 200 \times x + c$

Step 4: Calculate the constant, c

To work out c, we replace the x and y with the coordinates of either of our two points
$1000 = 200 \times 5 + c$
And rearrange c = 1000 − 200 × 5 = 0. So c is zero, as expected. This gives us our equation to work out the rough value of the area of a city from its population
$y = 200 \times x$

Step 5: Use with care

As with the line itself though, this equation should be treated with great caution—the values it gives are no more accurate than the line itself, and the accuracy of that can be judged by the

How it works

Lines of all kinds

Although so far we have been talking about straight lines, we can use exactly the same approach to find which curves fit most closely to points. We can find a line that passes through all four points below, for example—a perfect fit.

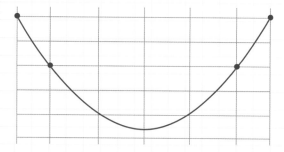

The general equation for this curve is $y = a + bx + cx^2$, and it is known as a quadratic. It is always possible to come up with a curve which passes exactly through any set of points; the problem is that the equation becomes increasingly complicated. These six points, for instance:

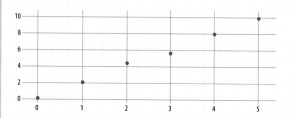

… can be perfectly fitted by the following curve:

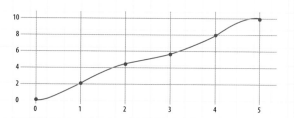

But the equation for this curve is
$$y = 0.1 + 0.2x + 2.9x^2 - 1.2x^3 - 0.01x^4 + 0.07x^5 - 0.01x^6$$

It's hard to imagine explaining where such a complicated relationship might come from. A more likely answer might be that the underlying relationship is something simpler. This line, for instance:

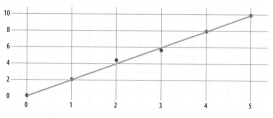

… which has the equation $y = 2x + 0.2$.

In practice of course, the points will come from somewhere and we should have an underlying theory to explain the relationship we expect to find. Usually, the points display data which has been measured, and since those measurements will not be perfect, there is no problem with a line which misses them slightly. The tricky thing is defining what "slightly" means! That topic is covered on page 90.

LINES WITHOUT MEANINGS

The great advantage of line-fitting techniques is that they will always give you a line (or more than one line) which is the best fit to any set of points. But this is also their great disadvantage, because the line might mean nothing: there may not be any relationship at all between the values you are plotting. You could plot the shoe sizes of a hundred people against their favorite numbers, and it would be possible to find the best-fitting straight line. The problem is that "best" here doesn't mean "good" or "meaningful." Of course, no one would think there is a relationship between how big your feet are and which number you like best, but what about shoe size and height? Would the best straight line mean anything in that case? This question goes to the heart of the point of statistics, which is to help test, prove, and develop theories. Unless there is some reason to think two pieces of data are related, there is no point in plotting them, or fitting them, either.

Knowing a person's shoe size has limited value.

scatter of the data points around it. A question that lies behind all of the above is: where do these lines come from? Although we could fit many lines by eye and then use least squares calculations to find the best fit, this is very time-consuming, and there is no way to be sure that there is an even closer-fitting line that has not yet been tried. There is a mathematical technique to find the line of best fit, called regression and explained on page 118.

SEE ALSO:
▶ Error, page 90
▶ Regression, page 118

Laplace's Demon

THE MAIN PURPOSE OF
STATISTICS IS TO TELL US
THINGS ABOUT AN UNCERTAIN
WORLD. If someone throws a
hundred dice in the air, statistics
could predict quite accurately
how they would fall—up to a
point. It would tell you
roughly how many 6s to
expect to see, but not
whether any particular
dice would fall as a 6.

The dice are cast. How will
they fall? Ask a statistician
such as the French genius
Pierre-Simon Laplace (above).

The way the dice fall depends on how they are
thrown, the air they fall through, and the ground
they land on. If some brilliant scientist knew
enough about these things, could that scientist
predict just which dice would fall as 6s?

The power of prediction

French mathematician Pierre-Simon Laplace
would have said yes, and scientists today would
agree. Given enough information, the fall of
individual dice could be predicted using the laws
of physics. But Laplace went much further and
imagined an "intellect" (now usually given the
name Laplace's Demon), which knew so much
about every atom in the Universe that it could
predict exactly what would happen at any point

in the future, in complete detail. While you may think that you can choose freely which cookie to take next, Laplace would say that the Demon, knowing exactly how your brain works, would be able to tell in advance which one you will select.

The statistics of violence

Laplace was a genius who made major contributions to many areas of science and mathematics. He developed improved versions of least squares techniques, Bayes' theorem (see more, page 52), and the central limit theorem. Even the French Revolution, which unfolded in the late 1780s while he was living in Paris, did not stop Laplace's research. (However, one of his closest and cleverest scientific colleagues, Antoine Lavoisier, was not so fortunate: he was sent to the guillotine in 1794. When his friends argued that, as France's (and the world's) greatest chemist, Lavoisier should be allowed to live, the judge's reaction was to say that "The Revolution has no need of scientists.")

The nebular hypothesis

Laplace's interest in statistics resulted from his greatest discovery: the origin of the Solar System. According to Laplace, the Solar System began as a vast cloud in space (called a nebula), which collapsed under the influence of its own gravity and began to spin. Rings of material were thrown

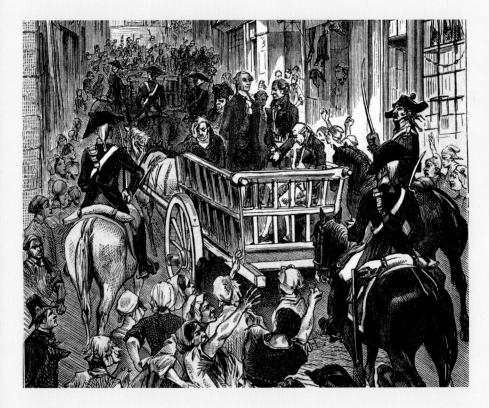

Antoine Lavoisier's immense intellect was not enough to save him from a date with the guillotine. Despite his many contributions to science, his downfall occurred because he had accrued enormous wealth by collecting taxes for the king from impoverished peasants.

off as the cloud contracted, and each ring then itself collapsed to form a planet. The core of the cloud shrank to a dense mass which became the Sun. This "nebular hypothesis" was more or less correct and is still used today.

Math of motion

Although at the time neither the mathematical techniques nor the data were available to make his theory a complete mathematical one, Laplace was able to develop further the mathematical theory of the Solar System developed by Isaac Newton in the 1660s. Newton had showed how to work out the motions of all the planets (and moons and comets) using just four mathematical laws of physics: three of motion and one of

gravitation. He explained this in his book "Mathematical Principles of Natural Philosophy," usually known by the first word of its Latin title, the *Principia*.

The fate of the Solar System

While Newton had been almost completely successful in explaining the motions of the planets, a question remained about whether the Solar System was stable. The planets pull on each other in ways that were (and are) impossible to calculate exactly, and it was not known whether these "perturbations" would eventually cause

Below: Newton's work was inspired in part by the motion of comets, especially the Great Comet of 1680.

Above: Some pages from Laplace's *Exposition du système du monde*.

Right: A clockwork model of the Solar System which can be used to predict the positions of the planets. Newton devised a mathematical model of the Solar System, but thought the System needed both a creator and a repairer. Laplace proved that it needed neither.

the Solar System to collapse, or perhaps to drift apart. Newton himself suggested that God might have to intervene occasionally, to avoid such disasters. Laplace proved that the perturbations would never become large enough to destabilize the Solar System, and that Newton's laws would continue to predict the motions of the planets for the rest of time. He published his conclusions in 1796 in his book *Exposition du système du monde*. When Napoleon Bonaparte, at that time the First Consul of France, discussed the book's findings

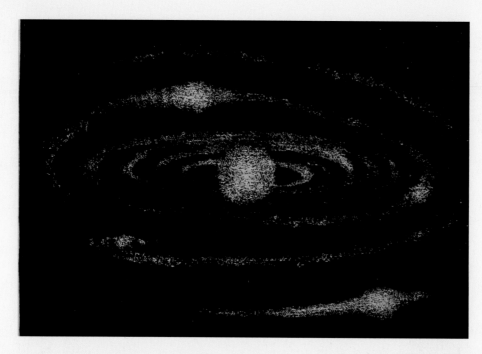

Laplace's theory of the formation of the Solar System was largely correct. This disc of dust and gas is collapsing to form a new star and planets, as Laplace suggested.

with Laplace, he asked why Laplace did not mention the influence of God. Laplace famously responded, "… I had no need of that hypothesis."

Information needed

Laplace's demonstration of the power of Newton's laws led him to conclude that absolutely everything is predictable with perfect accuracy, given enough information. It is the job of scientists to gather this information. But, since measuring instruments are not perfectly accurate, there will always be a gap between our measurements and the actual values needed to make perfect predictions. With Laplace's bold new belief, the study of this gap became crucial. If you like running, you might be quite happy with a watch that tells you your speed is 11 yards per second. But if someone calculates that your personal maximum speed is 11.689 yards

a second, you might have a lot of new questions about the accuracy of your watch and how you might improve it.

Minding the gap

It is the job of statistics to study the gap and, because of this, Laplace took the subject seriously in a way that other mathematicians—who still associated it with gambling—did not. This is why Laplace was the first scientist to provide proper estimates of how accurate his answers were. For instance, he calculated the mass of Saturn to be 1/3512 that of the Sun. Similar calculations had been made before, but what was new was that Laplace also calculated that the odds of his estimate being off by more than 1 percent of the true value were just 1 in 11,000. (The correct value is about 1/3499, which is indeed within 1 percent.)

COULD THE DEMON DO IT?

Modern science has shown that Laplace was incorrect; even a being who knew everything there is to know about the Universe could never predict its future exactly. This is because there are certain phenomena which are genuinely random and therefore unpredictable. For example, the nucleus of a radioactive atom may release a particle (see more, page 62) at any time and that time could never be known in advance. Secondly, making a perfect measurement is impossible, because every measuring instrument will very slightly (and unpredictably) change the thing that is being measured.

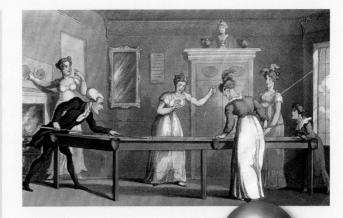

Laplace and many of his contemporaries believed that the Universe is composed of atoms which interact with each other according to Newton's laws, as pool balls do. So, just as we can predict the arrangement of pool balls after the next shot by using observations and mathematics, we should be able to predict the future arrangements of all the atoms in the Universe.

An anemometer measures wind speed, but the energy the wind loses to make it spin means the speed of the wind is slightly reduced. So, the true wind speed cannot be measured exactly.

SEE ALSO:
▶ What to Expect, page 28
▶ Randomness, page 112
▶ Measuring Confidence, page 154

Error

IN 1796, A 24-YEAR-OLD ASTRONOMER'S ASSISTANT CALLED DAVID KINNEBROOK WAS FIRED BY HIS BOSS, Nevil Maskelyne, Britain's "Astronomer Royal," for repeatedly getting the wrong answer when noting down his observations of stars. Three decades later, his mistakes proved to be very useful in understanding the mathematics of error.

Kinnebrook worked with Maskelyne at the great observatory in Greenwich, close to the center of London (which to this day is used as the location of the 0° prime meridian from which longitude is measured). Kinnebrook's failure had involved estimating the time at which a star passed a particular point in the field of a telescope, to an accuracy of a tenth of a second. The young assistant managed to find a job in a school for a while, and was later forgiven by Maskelyne, who found his ex-assistant a job as a "computer," calculating values for a nautical text book.

Left: Nevil Maskelyne.

Below: The Royal Observatory, Greenwich.

A kindred spirit

In 1820, the German astronomer Friedrich Wilhelm Bessel heard about Kinnebrook's dismissal and began to wonder just how the problem had arisen. Like Kinnebrook, Bessel had been an astronomer's assistant and a computer. From the age of 15 he had been apprenticed to a firm which made its money through import and export; as was usual then, his seven-year apprenticeship was an unpaid position, but he was such a skilled mathematician that, within a year, he was being paid to calculate the company's accounts.

The personal equation

Bessel gathered together similar data from other astronomers and, when he examined them statistically, he soon saw that each had his own personal time-delay in recording observations. (And it was "his" because there were practically no female astronomers at the time, though there were many female computers.) Bessel referred to this as the "personal equation" of the observer.

Until the 1940s, all "computers" were human beings. A few of the first prominent computers were Alexis-Claude Clairaut, Nicole-Reine Lepaute, and Joseph-Jérôme Lalande, who all worked in 1757. Seventy-five years before, in 1682, Edmond Halley had predicted that a comet would appear in the sky in the year 1758 or '59. The mathematical trio divided up the task of calculating exactly where in the sky it should appear—and found that Halley's Comet, as it became known, reached its closest position to the Sun in April 1759.

The human factor

At this time, accurate measurements were both necessary for and the result of the new technologies of the Industrial Revolution. The high standards of the instruments themselves meant that the chief source of error in making a measurement would often be the person using them. So, Bessel's new concept was very timely—and it also meant that people like Kinnebrook would no longer be regarded as lazy or inattentive. In fact Bessel's work resulted in a change in the use of the word "error" (at least, when used by scientists) from meaning "mistake" to "difference between the actual value and a measurement of it."

Thinking fast

The "personal equations" of Bessel are closely related to the idea of reaction time. If you drop a ruler through a friend's open hand, the length that passes through their fingers before they catch it is an easy measure of their reaction time—and personal to them.

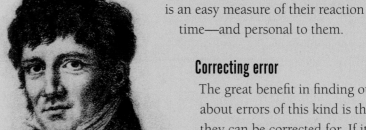

Above: Friedrich Wilhelm Bessel.

Correcting error

The great benefit in finding out about errors of this kind is that they can be corrected for. If it has always taken you about a tenth of a second to react to a signal, and you are in charge of manually timing some very rapid event using a stop watch, you could subtract one tenth of second from all your answers to make them more accurate.

What's your personal equation? Find out with this simple test.

How accurate is your stopwatch? To really find out, you will first need to know how much error you are adding as you use it.

Reading the temperature from this thermometer will depend on the exact position of the human viewer. However, they will all agree that it is a cold day.

Incorrect result

Neither of the two readings are exact, and if they were the times of two runners to complete a race, the first runner would be declared as the winner. In fact, they may be slightly slower than the other runner; and so the safest conclusion would be that the race was a dead heat because both results fall within the margin of error of a tenth of a second. This type of error, which is about the same each time an observation is made, is called a systematic error. Systematic errors also affect some machines. For example, the anemometer on page 89 is likely to lag a little behind the air that moves past it, and so will slightly under-report the wind speed. Systematic errors can be corrected for, at least to some extent.

Just a moment …

But correcting for errors like this can only be done roughly. Reaction times, like personal equations, are averages. It may take 0.08 second to press a stopwatch button on one occasion, and 0.19 second another time. What this means is that a stopwatch, which can record time to a hundredth of a second or less, should be used with caution. A person using it to time two things which, in actuality, take exactly the same time, will almost certainly produce two distinct readings, like 00:39:59.99 and 00:40:00.04.

Random errors

Rulers, glass thermometers, spring balances, rotary clocks, and other equipment in which a measurement is taken by judging the position of pointer or level, can only be read to a certain accuracy, so the measurements made will cluster around the actual value. These are random errors, and unlike systematic errors

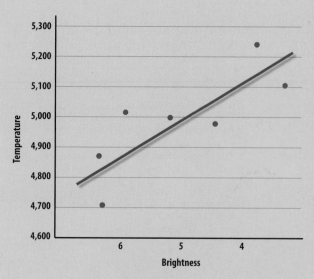

Estimated temperatures of stars compared to their brightnesses.

THE WINK OF AN EYE

One type of reaction time leads to a phenomenon called persistence of vision. The brain retains an image from the eye for a fraction of a second, and if the eye is presented with images that change more rapidly than this, they will be seen together at the same time. This is the basis of the "bird in the cage" illusion, and is why we see moving images on a cinema screen rather than a succession of still photographs, too.

they lie either side of the true value and cannot be corrected. Random errors can be expressed using the ± symbol (meaning "plus or minus"). A weight measurement written as 33±1 ounces means the size of the random errors is such that the actual weight may be anywhere from 32 to 34 ounces.

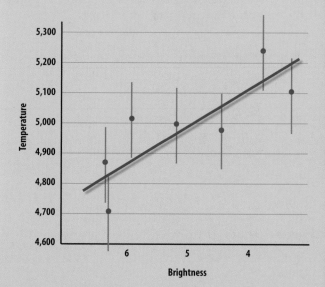

**Accurate
and precise**

**Not accurate
but precise**

**Fairly accurate but
not precise**

**Not accurate
and not precise**

Error bars

In analyzing the results of an experiment, we often try to fit a line through a plot of the results (see more, page 122), but the best-fit line frequently does not pass through all the points— and sometimes doesn't pass through any of them. The temperatures of stars as shown in the chart top left can be very hard to measure. If the errors in the values are only accurate to ±150 degrees, we can add error bars which extend 150° above and 150° below each point, as shown in the chart bottom left. Because the line passes through all the error-bars, it seems safe to say that it is a good fit to the observations.

Accuracy vs. precision

Although accuracy and precision are often treated as if they mean the same thing, accuracy is really about the quality of measurements, and precision is about how values are written down. If you can only measure temperature to the nearest degree, then this is less accurate than measuring it to the nearest one-tenth of a degree. On the other hand,

12.676 is a very precise value, 13 is less precise. Precise archers will cluster their arrows close together on the target, whereas archers who are both accurate and precise will hit the bulls-eye every time. These results can also be stated in terms of errors. Closely-clustered results have a small random error. If they are clustered but off target, then they are affected by a systematic error (that might be caused by a steady breeze). Arrows that are consistently close to the target, but not close to each other, are fairly accurate but not precise. Similarly, if you use a flat ruler to measure the circumference of a football, your answer is likely to be very inaccurate—perhaps accurate to about half an inch. So, the answer should be recorded only to the nearest half-inch. Writing it to the nearest 0.1 inch would be more precise, but misleading.

SEE ALSO:
▶ The Average Human, page 96
▶ Outliers, page 108
▶ Fallacies, page 168

The Average Human

ALTHOUGH SCIENCE IS THOUSANDS OF YEARS OLD, the idea that people can be scientifically studied just as planets or animals can be—and the idea that people are animals ourselves—is still quite new. Scientists, like other scholars, liked to believe they were better than the rest of the world, perhaps created in God's image and certainly not controlled by impulses like animals are. One person who helped changed all this was the mathematician Adolphe Quetelet.

Adolphe Quetelet, right, was born following a period of great upheaval in what is now the country of Belgium, which was rocked by violence in the 1780s and '90s.

Quetelet was born in Ghent in 1796 during a time of great political drama that followed the transfer of the city from Austria to Belgium and then back to Austria again in 1790, before it was taken over by France in 1795. Quetelet loved both math and the arts—he became a math teacher but also painted, wrote poetry, and even collaborated on an opera with Germinal Dandelin, a fellow mathematician. Quetelet said that "Some people did praise it, but that didn't stop Dandelin declaring, after the second performance, that he would be among the first to boo the play if anyone suggested repeating it." In 1823, Quetelet moved to Brussels to help set up a new observatory there. He found that the most important area of mathematics for a budding astronomer like himself to learn about was statistics, and he was soon lecturing in this as well as in astronomy and physics.

Revolutions

In 1830, revolution swept through Belgium, which made that a period of great uncertainty for Quetelet himself. The national upheaval made him consider whether changes in society might be understood—and even controlled—scientifically, and explained by natural laws, just as the motions of the planets he studied could be predicted

Violence returned to Quetelet's homeland in 1830 in the Belgian Revolution.

by the natural laws of motion and gravitation that Isaac Newton had defined. Quetelet knew from the work of Laplace (see page 84) and Poisson (pages 100 to 107) that statistics could be applied to human beings, and he began to wonder whether statistics might provide the tools he needed. As a painter, he had long been interested in ranges of human body measurements, but now he wondered whether non-physical attributes of humans might be measured, too, including the tendency to commit crimes.

The power of numbers

Quetelet soon realized that although it might be impossible to find out why one person committed a crime, it might still be possible to study the causes of crime by statistics, but only if enough information was available about large numbers of criminals, for comparison with non-criminals. In 1831, he published a short pamphlet about

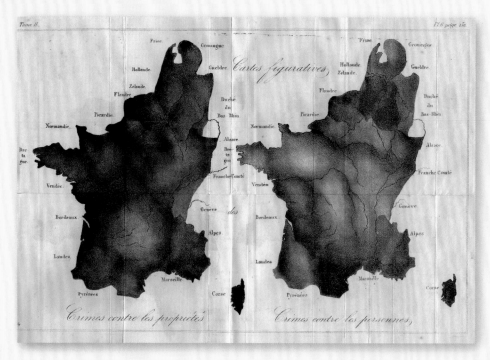

Tome II.   Cartes figuratives   176 page 150.

Crimes contre les propriétés,   Crimes contre les personnes,

Quetelet and his colleagues pioneered a new type of statistical map, in which darkness represents the number of crimes. In these 1831 maps, the left shows crimes against property, the right is crimes against people.

A criminal court in the days of Quetelet.

this topic, in which he said that "The greater the number of individuals, the more the influence of the individual will is effaced, being replaced by the series of general facts that depend on the general causes according to which society exists and maintains itself." Four years later, he published a large report on this same topic and made an even stronger statement of his view, "Society prepares the crime, and the guilty person is only the instrument by which it is executed."

Radical views

These ideas caused outrage in some readers. Since rates of crime were related to factors like age, wealth, education, gender, and even time of year, did that mean that criminals could not help committing crimes? If so, was it right to punish them, or should they be helped? This debate is still very active today. Nevertheless, in Quetelet's

time, even very young children could be tried and imprisoned. Today, it is recognized in law that people younger than a certain age (which varies from country to country) cannot be blamed or punished for their actions to the same extent as older wrong-doers.

Average human

In physics, one way to study how a large and complicated group of particles behaves is to consider just one typical one, as was done with the neutron in early atomic research (see more, page 62). So, Quetelet defined an "average man" to be used in calculations, who had average measurements and also average non-physical characteristics (which Quetelet described as "moral" characteristics), including tendencies for crime, suicide, and courage. Quetelet's insights caused a revolution of their own, in the way in which people can be analyzed. Although it was already known that groups of people behave according to their own rules, different from those of individuals, the idea that the behavior of groups could be analyzed scientifically to reveal new characteristics was fresh and exciting. When we use the word "society" today, what we mean is what Quetelet meant.

A final revolution

Long after his death, Quetelet's work caused another kind of revolution. In the 19th century many poor people did not have enough to eat, so being fat was a good thing—a sign of wealth. But, as medical knowledge increased, it gradually became clear that being fat was unhealthy. In

Where he walked, freedom grew...

Despite what this advert attests, insurance has always been a business based on figuring out the odds of bad events happening, and Quetelet's work proved invaluable.

the 1940s, financial institutions began to sell life insurance, and so the calculation of people's life expectancies, and the effect of health on them, became important. But how should healthy and unhealthy weights be defined? Using a person's actual weight did not work, because weight depends on height as well as on the amount of body fat. The measure that turned out to be best is what is now called BMI (Body Mass Index), which is based on the principle that normal body weight is proportional to the square of height. This relationship had been discovered by Quetelet, over a century before.

In the 1800s, if you were fat then life was good.

SEE ALSO:
▸ Matters of Life and Death, page 32
▸ Correlation, page 122

Poisson's Distribution

THE BINOMIAL DISTRIBUTION IS A POWERFUL WAY OF MODELING RANDOM EVENTS, from simple things like the probability of getting 2 sixes when throwing three dice, to complicated ones like choosing the most promising new drug to treat a disease. But there's a big problem.

Left: Siméon-Denis Poisson.

As we saw, the equation for a binomial probability is $P(x) = n!/(n-x)!x!p^x q^{(n-x)}$, and in it, q is the probability of the thing you want not happening. If you are throwing a dice and hoping for a three, then p = 1/6 and q = 5/6. But what about events like lightning strikes, or touchdowns in football? You can count lightning strikes per storm, or touchdowns per game, but how many times per hour does lighting not strike; how many touchdowns are not scored? Without those numbers, we can't use the binomial distribution. Instead, we use a formula devised by Siméon-Denis Poisson, who was born near Orléans in France in 1781.

Poisson falls in love

Poisson began his education by studying to be a surgeon, but turned out to be much too clumsy for the work. When he turned to mathematics, studying at the famous École Polytechnique in Paris, he found his vocation, saying that "Life is good for only

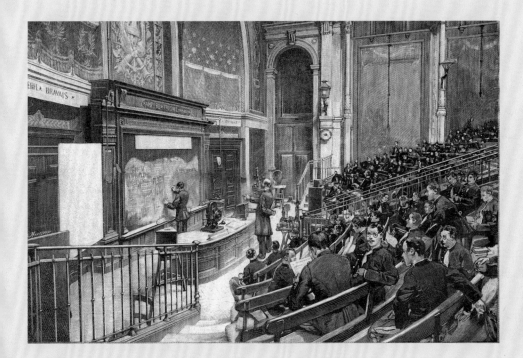

The École Polytechnique was and is the premier engineering and mathematics college in France.

two things, discovering mathematics and teaching mathematics." Poisson's teachers included the two greatest mathematicians of the time, Pierre-Simon Laplace and Joseph-Louis Lagrange. Because he was so extremely butter-fingered, not to say ham-handed, he couldn't draw the diagrams he needed to do well at geometry, but he excelled at mathematical physics and statistics. Although Poisson had no interest in politics, he did argue against his fellow students when they wanted to publish their criticisms of Emperor Napoleon in 1804. This was only because Poisson

Above: Joseph-Louis Lagrange.

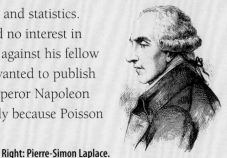

Right: Pierre-Simon Laplace.

was concerned it might disrupt the École's mathematics teaching, but Napoleon thought he had found a new supporter, and later helped Poisson join the newly-created Faculté des Sciences. Poisson wrote more than 300 scientific papers, and was one of the most well-organized of all great thinkers. To avoid working on more than one major breakthrough at the same time, he would make a note of each new brainwave and keep the notes in his wallet. When each new discovery was safely written up and ready for publication, he would go through the notes to choose his next world-changing project.

The last game uses a 26-sided dice, labeled with all the letters of the alphabet. You can throw it 26 times, and you win by throwing a Z. Which should you choose? There is a formula to work this out: luckily, in each case the number of rounds is the same as the number of possible outcomes. If we call this number **N**, then the probability of winning (let's call it **P(win)**) is

$$\left(1 - \frac{1}{N}\right)^N$$

So, for the coin game, where we have two rounds and two possible outcomes for each toss (a head or a tail),

$$P(win) = \left(1 - \frac{1}{2}\right)^2$$

which is **(1-0.5)² = 0.5² = 0.25**.

For the tetrahedron game, with its four rounds and four outcomes,

A tetrahedral dice has four sides so you can score from 1 to 4.

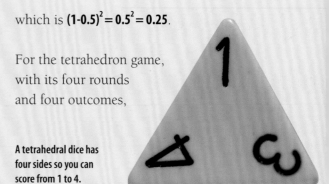

e

Poisson worked out how random events were distributed by using the number *e*, sometimes called Euler's constant. *e* lies behind many natural processes. There are many ways to derive *e*; this one is based on the probabilities of random events. Imagine you are challenged to play one of three games of chance. How do you decide which offers the best chance of winning? One game is very simple: you toss a coin twice, and you win by getting at least one head. The second game is played with a tetrahedron with red, green, blue, and yellow faces. You throw it four times, and you win if it falls at least once with the red side down.

This mnemonic helps you to remember the first 13 digits of *e*.

2.	7	1	8	2	8
to	express	e	remember	to	memorize

$$P(\text{win}) = \left(1 - \frac{1}{4}\right)^4 = (1-0.25)^4 = 0.75^4 \approx 0.3164$$

($\approx$ means "approximately.") For the letter game, the formula gives:

$$\left(1 - \frac{1}{26}\right)^{26} \approx (1-0.0385)^{26} \approx 0.3607$$

We can see that as N increases, P(win) rises, so the letter game is the best to choose.

From chance to e

Although P(win) is higher each time, it is not much more for the letter game than the tetrahedron one. If we calculate what happens as N increases further, we will find that P(win) hardly changes at all:

$$P(\text{win})_{(N=100)} = \left(1 - \frac{1}{100}\right)^{100} \approx 0.3660$$

$$P(\text{win})_{(N=1,000)} = \left(1 - \frac{1}{1,000}\right)^{1,000} \approx 0.3677$$

$$P(\text{win})_{(N=1,000,000)} = \left(1 - \frac{1}{1,000,000}\right)^{1,000,000} \approx 0.3679$$

The number which P(win) is approaching is about 0.36787945, and this number occurs many times in probability and elsewhere. Most often it is the reciprocal of this number (1/0.36787945)

that is used, which is about 2.718281758, symbolized by the letter e and known as Euler's number. e can be defined entirely in terms of factorials (see page 23):

$$e = \frac{1}{0!} + \frac{1}{1!} + \frac{1}{2!} + \frac{1}{3!} + \frac{1}{4!} + \cdots = \frac{1}{1} + \frac{1}{1} + \frac{1}{2} + \frac{1}{6} + \frac{1}{24} +$$

From e to Poisson

Deriving Poisson's distribution from e is quite complicated, but can be simplified by cheating. To start, we must find a formula for e raised to any power at all (let's call the power λ). The simplest way is to use the factorial formula from above, and raise that whole formula to the power λ:

$$e^\lambda = \left(\frac{1}{0!} + \frac{1}{1!} + \frac{1}{2!} + \frac{1}{3!} + \frac{1}{4!} + \cdots\right)^\lambda = \frac{\lambda^0}{0!} + \frac{\lambda^1}{1!} + \frac{\lambda^2}{2!} + \frac{\lambda^3}{3!} + \frac{\lambda^4}{4!} + \cdots$$

Now, we use the fact that any probability distribution adds up to 1. We are going to use e^λ as the basis of our new probability distribution, so we must make sure e^λ equals 1. Unfortunately though, our formula for e^λ goes on forever, so this may be tricky. The answer is to cheat, by doing something which will only make sense after we've done it. So, we say that in any fraction at all, n/n=1, no matter what n is.

Let's make $n = e^\lambda$. So, $\dfrac{e^\lambda}{e^\lambda} = 1$

1	8	2	8	4	5	9
a	sentence	to	simplify	this	tough	operation

Feeding in the equation above to replace one of these e^λs gives

$$\frac{e^\lambda}{e^\lambda} = \frac{(\frac{\lambda^0}{0!} + \frac{\lambda^1}{1!} + \frac{\lambda^2}{2!} + \frac{\lambda^3}{3!} + \frac{\lambda^4}{4!} + \cdots)}{e^\lambda} = 1$$

Now we simplify

$$\frac{\lambda^0}{0!\,e^\lambda} + \frac{\lambda^1}{1!\,e^\lambda} + \frac{\lambda^2}{2!\,e^\lambda} + \frac{\lambda^3}{3!\,e^\lambda} + \frac{\lambda^4}{4!\,e^\lambda} + \cdots = 1$$

And this series defines the Poisson distribution.

Poisson probabilities

λ equals the mean number of events that we are interested in, like the average number of lightning strikes per storm. The terms of the series tell us the probabilities of different numbers of strikes. The first term, with the zeroes, refers to the probability of an event not occurring. For instance, if the average number of strikes per storm is three, then the probability of a storm with no lightning is calculated by changing the λ in the first term to a 3, to give

$$P(0) = \frac{3^0}{1 \times e^3} = \frac{1}{(2.718281758\ldots)^3} \approx \frac{1}{20} = 0.05 = 5\%$$

Number of events	0	1	2	3	4	n
Probability of that number of events	$\dfrac{\lambda^0}{0!\,e^\lambda}$	$\dfrac{\lambda^1}{1!\,e^\lambda}$	$\dfrac{\lambda^2}{2!\,e^\lambda}$	$\dfrac{\lambda^3}{3!\,e^\lambda}$	$\dfrac{\lambda^4}{4!\,e^\lambda}$	$\dfrac{\lambda^n}{n!\,e^\lambda}$

We can summarize the above series as

$$P(x) = \frac{\lambda^x}{x!\,e^\lambda}$$

In text books, this is usually written as

$$P(x) = \frac{\lambda^x e^{-\lambda}}{x!}$$

Letters to numbers

Now all we have to do is work out the values of the terms in the series. Let's say we study 1,000 storms and find that there are three lightning strikes per storm on average. So e^z becomes e^3, which is about 20. So, we can replace e^z in the formula with 20.

$$\frac{1}{1 \times 20} + \frac{3}{1 \times 20} + \frac{9}{2 \times 20} + \frac{27}{6 \times 20} + \frac{81}{24 \times 20} + \cdots = 1$$

This gives a final set of values, of about

$$0.05 + 0.15 + 0.225 + 0.225 + 0.169 + \cdots = 1$$

These are the probabilities of having 0, 1, 2, 3, and 4 strikes per storm. Over our 1,000 storms, that means we should expect about 50 with no strikes, 150 with one strike, 225 with two strikes, about the same number with three strikes, 169 with four, and the remainder (that is, about 181 storms) with more than four strikes. Plotting the probabilities (and extending the calculation further to give the probabilities of higher numbers of strikes) gives the shape of a Poisson distribution (shown near right).

ON TARGET

The Poisson distribution can give results of amazing accuracy. In World War II, London was attacked by flying bombs, which were unmanned planes, filled with explosives, that flew until their fuel ran out, at which point they crashed and exploded. Just how clever were these bombs? Were they falling randomly? Or were they being aimed and timed to hit particular targets? To find out, the number of hits in 576 equal-sized London districts were recorded and compared with a Poisson distribution. The results showed very clearly that the bombs were falling in a random pattern.

Number of hits n	0	1	2	3	4	> 4
Number of districts with n hits	229	211	93	35	7	1
Poisson prediction	226.7	211.4	98.6	30.6	7.1	1.6

Unlike the normal and binomial distributions, the Poisson distribution is not symmetrical, and its shape changes depending on its values.

The chart below right shows the Poisson distributions for $\lambda = 1$ (blue), 3 (red), and 5 (green).

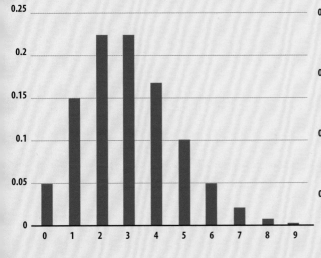

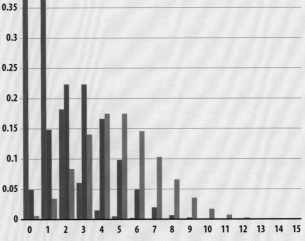

WHOA!

An early use of the Poisson distribution became famous in 1898 when Ladislaus Bortkiewicz, a Russian statistician who was also a colonel in the cavalry, applied it to the numbers of deaths by horse kicks in the Prussian Army. Below are the numbers of deaths caused by a horse kick over 20 years.

Year	Deaths
1875	3
1876	5
1877	7
1878	9
1879	10
1880	18
1881	6
1882	14
1883	11
1884	9
1885	5
1886	11
1887	15
1888	6
1889	11
1890	17
1891	12
1892	15
1893	8
1894	4

At first glance, some of these numbers may seem a bit odd. Why was 1880 such a bad year? Why is there a run of bad years from 1889 to 1892? Why do the lowest number of deaths occur in the first and last years? These questions all arise because of our natural human tendency to look for patterns and meaning in data. Perhaps these questions might have led a less mathematically minded colonel to make investigations and inquiries about the good and bad years, but Bortkiewicz was able to answer them all without leaving his office. He considered separately each of 14 cavalry units, noting the annual deaths in each over a 20-year period, giving him 14 x 20 = 280 combinations.

He then counted all the combinations in which there were no deaths, 1 death, and so on, and compared the result (shown below in blue) with the predictions of a Poisson distribution (in yellow).

As the chart shows, the distribution is a completely random one, demonstrating how our instincts can lead us to the wrong trees and make us bark, when statisticians are not there to put us right. Bortkiewicz also applied the Poisson distribution to suicide rates and found that they, too, are random (for the times and places he chose). So, like Quetelet (see more, page 96), he showed that even things which seem to happen for very special and unique reasons, like suicides or murders, are in some ways random, which leads to questions about just how much control people really have over their actions.

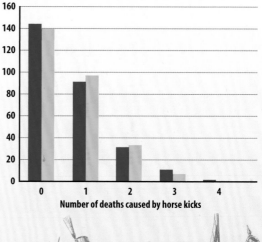

Number of deaths caused by horse kicks

Poisson's dots

Recognizing random patterns is something else we are naturally poor at. Which of these two patterns shown at the top of the page do you think is more random, resembling the sort of pattern that hailstones make on the ground or snowflakes in the air? Or are they both equally random? Thanks to Poisson, we can answer this question. We add grids to these two patterns, count the number of dots per grid-square, and plot the distribution in each case. The lower boxes show what we get. (The small charts to the right of each pattern show the distributions.) The left pattern has one dot per square, which is highly unlikely, but the right one closely approximates a Poisson distribution.

This kind of pattern analysis is used in medical and forensic laboratories. Some of the cells and creatures in blood, for example, form groups, and information about these groups is important in diagnosing diseases. A hemocytometer is a slide with a grid on it, which is examined under a microscope. The numbers of cells in each square are counted, and the numbers compared with a Poisson distribution. In this way, any groupings not due to chance can be identified. Poisson distributions are also used to analyze the flow of traffic by predicting the arrival times of vehicles, and to anticipate the pattern in which email messages arrive at a computer, to ensure that it has the capacity to cope with randomly busy periods.

SEE ALSO:
▶ Non-Parametric Statistics, page 140

Outliers

CHENOPIS ATRATA.

The Black Swan of Australia.

The chance of seeing a black swan was once regarded as very low—until Europeans arrived in Australia, where swans are black, not white.

IN THE NORMAL DISTRIBUTION AND MOST OTHERS, the probability drops away the further from the center you get. But it never quite falls to zero. This reflects the reality that any event might possibly happen: throwing six dice and getting 6 sixes is very unlikely, but not impossible.

Statistics is all about solving real-world problems. If someone really did throw 6 sixes, you are likely to wonder whether there is something unusual about the dice, or the thrower. So, the question is, how unlikely does something have to be before we decide that the event has not arisen by chance?

Samples and populations

Unusual results occur for many reasons. If someone catches the flu and is better two days later, it may be that they have a much better immune system than average. But it may also be that they never had flu in the first place, just a regular cold. In statistical terms, this means that they may not actually have been a member of the population of interest (the "people with flu" population). Gathering data from the wrong population can be a serious problem for surveys. Let's say you wanted to find out how tall schoolchildren are, so you leave

Figuring out how tall children are is a growing business!

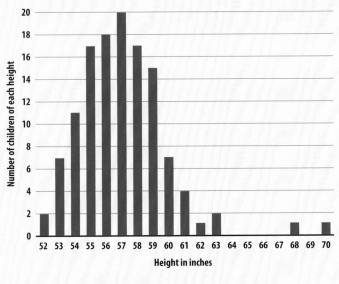

for the school, then it will make a big difference if you need to order chairs big enough to accommodate the tallest children. Secondly, the outliers might be errors. Maybe those two people measured themselves wrongly, or wrote the wrong height,

a pile of forms in a school for people to fill in. You get the results shown above. The two people who are 68 inches and 70 inches tall seem suspiciously different to the majority, and are called outliers.

What's the story?

There are at least three possible reasons for these outliers. First, there may indeed be two children in the school who are much taller than the rest. If there are, and if your reason for doing the survey is because you need to order new chairs

Spot the outliers here.

PEIRCE SAVES THE DAY

Benjamin Peirce was a famously calm man. He was a lecturer at Harvard University for fifty years, and every day he would just write steadily on his blackboard without pausing, explaining, or hardly looking at his class. One day in 1858, Peirce's calmness was a lifesaver. He was attending a concert in Fitchburg, Massachusetts, which had been organized by Phineas T. Barnum, a showman famous for putting on spectacular entertainments. This concert starred a Swedish singer called Jenny Lind, who was very famous indeed thanks partly to Barnum's advertising skill. This was her final concert in America, and Barnum had sold all the tickets he could—far more than the number of people the hall could hold. Barnum had booked the biggest hall he could find, rather than the best, and it was really just a big room over a railway station. Steam trains passed beneath the flimsy floor every few minutes. As the performance began, more and more people crushed in, temperatures and tempers rose, and the floor began to sag and creak. When people started to break the windows to cool the air, a panic began, and there would have been a riot if Peirce had not made his way to a table and climbed up on it. He stood there, motionless and unspeaking, and gradually the panic died away until he could be heard. He then suggested that if everyone kept still and quiet, all would be well. And so they were. The concert ended up being a great success.

Above: Jenny Lind in action.

or perhaps you misread their forms, or typed them in wrongly. The third possibility is that the data is entirely correct, but two teachers have picked up the forms and filled them in as instructed. In that case, you have sampled two populations (adults and children) by mistake.

Dealing with outliers

There are three ways to deal with outliers.
1) Check the data: look for errors in typing and reading, and confirm that there is no contamination by another population.
2) Choose a new statistic. If you are interested in

the central value, then the simplest solution is to avoid using the mean, and instead use the mode or median. For instance, the mean of the numbers 16, 17, 18, 19, 19, 19, 19, 20, 21, 21, 31, is 20, but if the 31 is rejected as an outlier, the mean drops to 18.9. Whether the 31 is included or not however, the median of the data is 19 and so is the mode.

3) Use a statistical test. Unfortunately there are many to choose from, all are quite complicated, and they give different results—which goes to show just how slippery outliers can be.

The Peirce criterion

One of the most popular and simple tests for outliers was invented in 1852 by Benjamin Peirce, an American statistician (see more about him in the box opposite.) This is how it works. First, calculate the mean and standard deviation of the data. For the above data, these are 20 and about 3.77. Subtract the suspicious value from the mean. If the answer is negative, ignore the minus sign. For our data, this is 20–30 = –10, so the number we want is 10. Divide this by the standard deviation 10/3.77 ≈ 2.66. The next stage is to calculate what Peirce called an R-value. This is very complicated to work

out, but luckily tables are available. Part of one is shown below left. We look up R for the size of our sample, which is 11, giving an R-value of 1.925. If our calculated value for the suspicious number is larger than this, which it is, that means that the suspicious number is an outlier and can be safely be rejected. Probably.

Benjamin Peirce, standing, is in discussion with Louis Agassiz, the geologist who identified the ice ages.

Number of observations	R-value
3	1.196
4	1.383
5	1.509
6	1.610
7	1.693
8	1.763
9	1.824
10	1.878
11	1.925
12	1.969
13	2.007
14	2.043

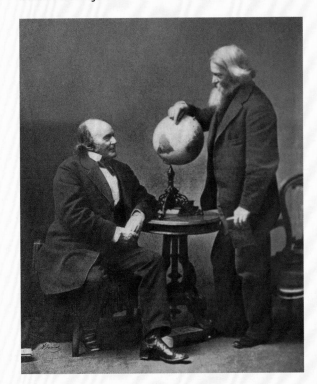

SEE ALSO:
▶ Correlation, page 122
▶ Compare and Contrast, page 146

Randomness

RANDOMNESS VERY OFTEN MATTERS IN STATISTICS. It is essential that surveys are random, for example. And, if we are looking for a connection between two groups of things (such as human heights and weights), we need to know what the data look like when they are not connected, which can be difficult.

Randomness matters to us all. If you play a lottery, or bet on the fall of a coin, or throw dice to determine the next move in a game, you rely on the randomness of the outcome. Random numbers are essential, too, in encoding information, so that financial transactions over the Internet are secure. Even defining randomness properly is difficult, and, as with a lot of commonly used words which are also mathematical terms, "random" to a statistician may not mean the same to anyone else. But there are two key features for a statistician: a random sequence has no pattern, and random things are unpredictable.

Patternlessness

Patterns in data can be easy to see, and a great deal of statistics has been developed to help us identify patterns. But how can you see patternlessness? Does this series have a pattern?

9, 7, 0, 2, 3, 1, 6, 4, 8, 5, 3, 1, 9, 7, 5, 8, 6, 4, 2, 0, 1, 8, 6, 3, 5, 7, 4, 0, 9, 2

Although each number is unrelated to its neighbors, there are two patterns here. Each number appears exactly 3 times, and each set of ten numbers contains the digits 0 to 9. There is actually no known test for patternlessness.

Vintage U.S. lottery tickets.

MARTIANS?

In the late 19th century, most astronomers thought that there was an advanced civilization on Mars, because they thought they could see a planetary network of purpose-built canals through their telescopes. In fact, there are no such things; even the highly-trained astronomers had been fooled by the tendency of the eye to connect random dots, patches, and blurs together to see lines that don't actually exist.

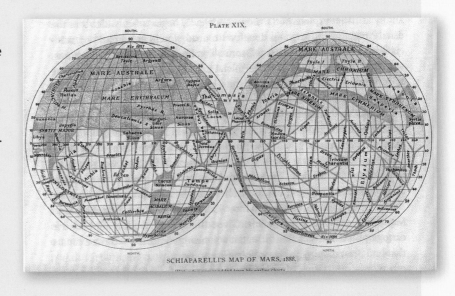

SCHIAPARELLI'S MAP OF MARS, 1888.

Unpredictability

There is no test for unpredictability either. Can you predict the next number in this sequence? 1, 5, 9, 2, 6, 5, 3, 5. The answer is 9: these are digits in the value of pi (π, the ratio of the circumference of a circle to its diameter), and 9 is the next one. However, pi is a transcendental number, meaning there is no pattern to its infinite string of digits.

Computerized randomness

Since almost all mathematical processes today are carried out by software, random number generation is an important feature of most computer systems. But there is no mathematical expression which can generate a random number, and that means that randomness must somehow be fed into the computer from the real world. This is not as simple as it might seem. In 1957 a machine called ERNIE (a convenient abbreviation of Electronic Random Number Indicator Equipment) became famous in the UK as the heart of the premium bonds system, which was a way of saving money by buying government bonds, with the incentive of regular bonus prizes. The prize winners were selected by randomly selecting the numbers of their specific bonds. ERNIE was the machine that generated the numbers, and it cost the equivalent of about £600,000 ($760,000) in today's money. Although it was often referred to as a computer and depicted as being super intelligent, in fact, random number generation was the only thing ERNIE could do. And it took 52 days to do it.

Pseudo-random

Today, there are two kinds of computer-based random numbers: Pseudo-Random Number Generators and True Random Number Generators. Pseudo-Random Number Generators (PRNGs) are usually based on stored lists of numbers; they are computerized versions of books of random numbers that used to be produced from about 1890 to about 1950. The key feature of a PRNG is that the long sequence of random numbers it produces can be repeated. This is often essential for software development: if the clouds in your interactive video game are generated randomly, then, while you are developing the game, you want to make sure they behave just the same way each time you use it. One problem with PRNGs is that they behave slightly differently in different operating

Above: ERNIE has been awarding random prizes in the UK since 1957. This is ERNIE 2 from 1972, while today's ERNIE 5 is 21,000 times faster.

Below: The unpredictable shapes created by lava lamps create one method used to generate random numbers for encrypting Internet traffic.

systems. The web programming language PHP, for example, generates very acceptable random numbers if it is used in a GNU/Linux system. But if it is used in Microsoft Windows, it produces lists of numbers with a distinct pattern.

Truly random

True Random Number Generators (TRNGs) usually use some kind of physical process to generate random signals, which are then converted into numbers. ERNIE used the amplified crackling sounds made from neon lamps, and some modern versions use radioactive chemicals, which produce unpredictable bursts of energy. TRNGs are used for those applications where the outcome must be impossible to know in advance—like games of chance. If only a small amount of random data needs to be generated, a completely physical process can be used, such as in lotteries or the draws in sports tournaments. Not only must the outcome be random, but it is just as important that people can see that the method is random, so, avoiding the use of computers in the process is an advantage.

Galileo's dice

The oldest way of finding random numbers is by throwing dice, and, apart from the possibility of weighted dice, or a thrower who cheats, a single dice will reliably generate a random number between 1 and 6. To obtain a wider range of random numbers, multiple dice can be thrown, taking into account the different probabilities of getting different total scores: the top and bottom scores are least likely to occur, because there is only one combination of dice that can produce

Dice is the world's oldest game of chance and is based on entirely random rolls.

How it works

Logarithms

Logarithms turn up in many areas of science and mathematics. This is how they work: 10 squared, which is also called "10 to the power 2" can be written 10^2, and equals 100; 10 cubed ("10 to the power 3"), which is 10^3, equals 1,000. We can also find values for 10 raised to other powers. For instance, $10^{0.30103} = 2$. In all these sums, the power to which 10 is being raised is called an exponent, or logarithm. So, the logarithm of 2 is 0.30103, the logarithm of 100 is 2 and the logarithm of 1,000 is 3. Most calculators and phones have a logarithm function. Sometimes, it is some number other than 10 that is raised to a power to form a logarithm, although 10 is the most common. These numbers are called bases, so, for clarity, you may find "log to the base 10" or $\log_{10}$ used instead.

Logarithm is frequently abbreviated to log, but do not get confused between working with logs and logs. This illustration will help you identify tools needed for both types.

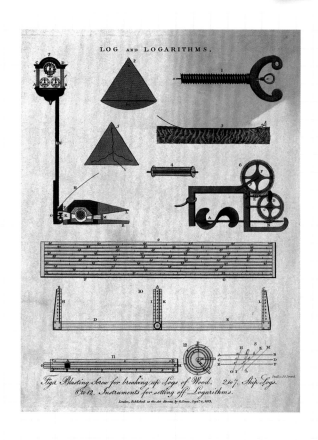

them. Unfortunately, there is an easy mistake to make in working out these probabilities, as the great Italian scientist Galileo Galilei was the first to explain, in 1576. In his time (and maybe today), most people would have said that the chances of scoring 9 with 3 dice was the same as scoring 10, because there are 6 ways in which 3 dice can add up to 9:

6+2+1; 5+3+1; 5+2+2; 4+4+1; 4+3+2; 3+3+3 and also 6 ways to get a total of 10:
6+3+1; 6+2+2; 5+4+1; 5+3+2; 4+4+2; 4+3+3. But actually there are far more ways to achieve a 9 or a 10. This is clear if we use different colors to represent the three different dice. Then we can see that, for example, 6, 2, 1 can arise in six different ways:

Are these real logarithm tables? Benford's law will not help because the numbers are not random.

1+2+6; 1+6+2; 2+1+6; 2+6+1; 6+1+2; 6+2+1.

This means that there are actually 27 ways to roll a 10, but 25 ways to roll a 9. So 10 is slightly more likely.

Real data?

Throwing dice is slow, very hard to automate, and tricky to feed into a computer efficiently. An easier way is to use some data which is already available and which should be random. For instance, there are many millions of electricity bills online, and one very tempting source of random numbers could be simply to use the number with which each meter reading begins. Surely, about as many readings will start with a 1 as with a 2 or a 9, so that should give ten different completely random numbers? If more are needed, we could use the first two or three digits instead. Sadly, this is very far from being the truth: amounts are much more likely to start with a 1 than a 9. This is called Benford's law after the American physicist Frank Benford who analyzed it in 1938, but it was first noted by his fellow American Simon Newcomb in 1881. While using a book of tables of logarithms (see box, left), he noticed that the first pages were dirtier than the last, which meant that the other mathematicians who used the book must have used numbers beginning with 1 much more often than numbers starting with 9. Benford's law

applies to population sizes, house prices, birth rates, areas of countries, and many other kinds of data. Because financial and economic data often follows Benford's law, it has been used to identify tax-dodgers: when people report their earnings and costs honestly, the numbers follow Benford's law, but if they simply make up the figures, the law does not apply.

SEE ALSO:
▶ Combinations and Permutations, page 20
▶ Fallacies, page 168

How, but not why

Although we can see that this method works, it's not clear why. Francis Galton, who invented it, did not know himself. Galton combined an interest in life sciences with a love of numbers, and trained in medicine for a while before abandoning it to study mathematics at Cambridge University. His father's death in 1844 occurred at just the right time for Galton. Already skilled in mathematics, he suddenly had the wealth to do whatever he liked, which was to study a field that we would now characterize as human geography. He traveled widely in Africa and Asia, observing, studying, and measuring the people he met there. (Galton also tried his hand at travel writing and at weather forecasting, producing an early form of weather map in 1875.)

The riddle of inheritance

Galton wondered how the physical differences he found between people from different parts of the world remained the same over many generations. Why was it that children resembled their parents so closely? What exactly was passed on from one generation to the next? For a while Galton thought that fingerprints might hold the key, and he developed the system

of fingerprint classification that is still in use today. He also tried to define the faces of different human "types" (like criminals) by merging individual photographs. But, for his purposes, both subjects were dead ends. Many scientists were puzzling over the mysteries of inheritance. Galton's uncle was the biologist Charles Darwin, who had worked out the process of evolution, publishing his theory in 1859. Evolution depends on children inheriting characteristics from their parents. Now we know that these features are transmitted by a complex chemical called DNA, which is present in all the cells of our bodies. However, the technology of the 19th century still had far to go to unravel the function of DNA molecules.

Galton's sketches pick out the common features of fingerprints that are still used to identify people.

Two routes to change

Galton appreciated that his question was complicated by the fact that not all characteristics are passed on biologically from parents to children. Language, for instance, is not, and for some characteristics, such as intelligence or courage, it is very hard to decide. Galton coined the phrase "nature or nurture" to express this fundamental question. Partly because of this interest, and partly because he just loved data, Galton collected a vast number of observations, including information about people who fidgeted during lectures and how bad tempered his friends were. He was very aware that he would not live long enough to observe everything he wanted to, and, like Buffon (see more, page 58), he was determined to avoid sleeping too much. He even invented a device to keep himself awake, called a "gumption reviver." It was designed to drip water on his head just as he dropped off. (He invented glasses for reading newspapers underwater, too, but did not invent a waterproof newspaper.)

Children of peas

To avoid the nature–nurture problem, in 1880 Galton began to study a case in which he was certain that characteristics were passed on biologically, by comparing the sizes of sweet pea seeds with the sizes of those of the parent plants. This began his development of the theory of regression, when he tried to prove if there is a link (called a correlation) between sets of data. Eventually, his techniques became very sophisticated, though he was always much more interested in statistics as a tool than as an object of study in itself. In fact, he made his most advanced statistical inventions in order to work out whether people's prayers are effective. (He proved that they are not.) Because of Galton's rough-and-ready approach to statistics, he tended to develop particular techniques for each new problem and, in fact, if he had stood back from his data and considered what they all had in common, he would no doubt have developed the powerful technique now called correlation. Nowadays, no one would attempt to analyze regression without

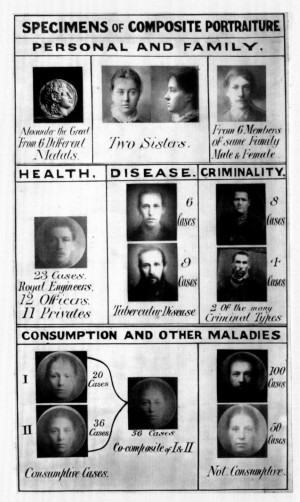

Galton was interested in linking physical characteristics, most often face and head shape, with intangible features like health and morality.

understanding correlation first, but that concept was not properly defined until several years later, by Karl Pearson (see more, page 122).

SEE ALSO:
▶ The Shapes of Data, page 38
▶ How Long Will You Live?, page 34

Correlation

FRANCIS GALTON'S REGRESSION METHOD OF FITTING THE BEST STRAIGHT LINE TO A SET OF DATA is a very powerful one, but you don't have to use it on many different data sets before you start to think that there is something not quite right about it.

Take the three graphs shown here. In each case, the regression formula has found the best straight line fit. All very clever and neat, but we usually do statistics for some reason, and line-fitting is generally done to find the equations that relate two variables together. Let's say that in these three graphs, the horizontal axis measures the dosage

of three experimental, new potato fertilizers, Alphagen, Betaphen, and Gammaphate, and the vertical axis is the extra crop yield in terms of percentage of weight. The first graph shows that Alphagen is doing the trick. Betaphen is also looking good, but perhaps is not as impressive. Gammaphate doesn't seem to work at all. Yet we have found a line in each case. So, can we do more? Can we measure the strength of the relationship between fertilizer dose and yield?

Measuring relationship strength

We can do this by measuring its correlation. This is measured on a scale that goes up to 1 (or 100 percent). A perfect correlation, like that between

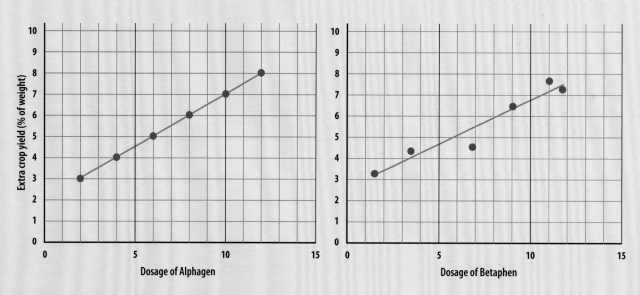

the number of dollar coins in a pile and the weight of that pile, has a value of 100 percent, but most correlations are weaker than this. Taller people tend to be heavier, so we can say that height and weight are correlated, but this is not a very reliable rule, so we say that the correlation is weak; perhaps 50 percent. When two things are unrelated, like the shape and color of fruit, there is 0 percent correlation. Correlations can be negative, too, like the relationship between the time to run a race and the speed of the runner.

Perfect correlation

It was Francis Galton who came up with the idea of correlation (see more, page 118), but was not able to calculate it properly. However, when he mentioned it in a lecture, one of the

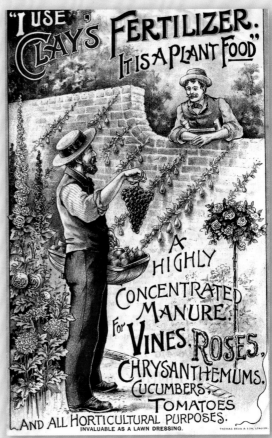

Which fertilizer works best? Such questions sow the seeds of a powerful statistical tool.

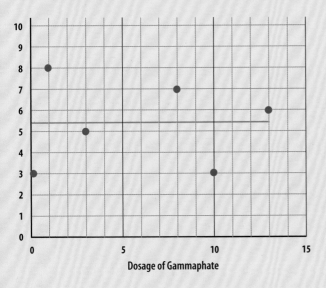

Dosage of Gammaphate

audience members was enthralled by the idea, saying that it "brought psychology, anthropology, medicine, and sociology in large parts into the field of mathematical treatment." The enthusiastic listener's name was Karl Pearson, and he was so inspired by correlation that he developed a whole new area of statistics based on it.

Pearson's coefficient

There are now a number of ways to calculate correlation, but the most popular is still the formula invented by Pearson in 1896, which is

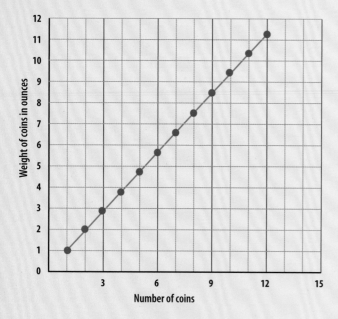

An example of a perfect correlation.

As with many other formulas in statistics, there are slightly different versions of this one for use with different kinds of data.

A simple example

Applying the PMCC involves many small calculations, and is only reliable when applied to large data sets (more than 25 points) and so it is usually calculated using software. But, to see how it works, we can use the data from page 119 to produce the table opposite. To keep track of the numbers, we can label the individual x and y values. The value reached, 0.992, is a high one, and

called Pearson's product–moment correlation coefficient (PMCC). There are different versions of the formula, but to see what it means, this may be the clearest:

$$PMCC = \frac{C_{xy}}{SD_x \times SD_y}$$

SD is standard deviation (page 68) and C_{xy} is called the covariance of x and y, which is defined as:

$$C_{xy} = \frac{\Sigma(x - \bar{x}) \times (y - \bar{y})}{n - 1},$$

where the "$\bar{\ }$" symbol means "mean", and **n** is the number of points.

Karl Pearson, left, visits with his mentor Francis Galton the year before Galton died.

this is consistent with the fact that the line passes very close to all three points (see chart, right). The reason why the coefficient should not be calculated with small data sets is that, if you scatter a small number of points randomly, they will normally form a rough line, and hence give a positive correlation coefficient even though there is no meaning in it.

Correlation and causation

Smoke is strongly correlated with forest fires, and obesity is strongly correlated with time spent watching television. But these correlations mean very different things. Forest fires cause smoke, but watching television does not cause obesity.

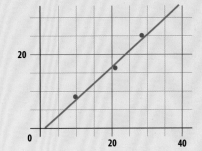

Wind speed (x)	Wave height (y)	
$x_1 = 10$	$y_1 = 8$	
$x_2 = 21$	$y_2 = 16$	
$x_3 = 29$	$y_3 = 25$	
$\bar{x} = 20$	$\bar{y} = 16.33$	
$(x_1 - \bar{x}) = 10 - 20 = -10$	$(y_1 - \bar{y}) = 8 - 16.33 = -8.33$	$(x_1 - \bar{x}) \times (y_1 - \bar{y}) = 83.3$
$(x_2 - \bar{x}) = 21 - 20 = 1$	$(y_2 - \bar{y}) = 16 - 16.33 = -0.33$	$(x_2 - \bar{x}) \times (y_2 - \bar{y}) = -0.33$
$(x_3 - \bar{x}) = 29 - 20 = 9$	$(y_3 - \bar{y}) = 25 - 16.33 = 8.67$	$(x_3 - \bar{x}) \times (y_3 - \bar{y}) = 78$
		$C_{xy} = \dfrac{\Sigma (x - \bar{x}) \times (y - \bar{y})}{2} = 80.5$
$SD_x = \sqrt{\left(\dfrac{\Sigma (x - \bar{x})^2}{n-1} \right)} \approx 9.539$	$SD_y = \sqrt{\left(\dfrac{\Sigma (y - \bar{y})^2}{n-1} \right)} \approx 8.505$	$SD_x \times SD_y = 81.13$

$$PMCC = \frac{C_{xy}}{SD_x \times SD_y} = \frac{80.5}{81.13} = 0.992$$

JUST A MOMENT

The name of the product–moment formula is hard to remember because the word "moment" is now hardly ever used in the way Pearson meant. In his time, four statistical "moments" were recognized. The first moment is the mean, and the second is the standard deviation. The others are skew (see more, page 132) and kurtosis (see more, page 145).

One explanation for the strong correlation between TV and obesity is that people who exercise are unlikely to spend much time watching TV and are also unlikely to be obese. Ice-cream sales, wearing of sunglasses, sunburn, and shark attacks are all correlated too, but none causes any of the others. So, just because two

Where there is smoke there is fire. Time to get out of there!

Quinine drinks and cases of malaria are closely correlated. Should quinine be banned?

things are correlated, that does not mean that one causes the other. If the actual data that is being studied is simply two sets of numbers, it is never possible to tell whether there is a causal link between them.

Cause or cure?

It's not always simple to work out how causality works. The amount of water drunk by people in developed countries is strongly correlated with being healthy. But is this because health-conscious people drink lots of water because they think it's healthy, or are they healthy because they drink lots of water? Most people in developed countries, healthy or not, have enough water to drink in any case, and it's not at all clear whether drinking extra water (as health-conscious people tend to do) makes them any healthier. Sometimes correlation will tell the opposite of the truth when used as a guide to causation. In the 19th century, people from the UK and other European countries and from the United States who lived in African and Asian countries, often consumed a great deal of quinine, which is a natural drug obtained from cinchona trees, and used to flavor a soft drink called tonic water. There is a very strong correlation between areas where a lot of quinine was consumed and those where there were a great many cases of malaria. Looking at the data on its own, it would be easy

to conclude that quinine should be banned. In fact, quinine was known to be very effective at tackling the symptoms of malaria, which is why it was so popular in malaria-stricken regions, so banning it would have been a very bad idea.

Comparing preferences

Correlation is a powerful tool, but it only works for numerical data. It can't be used, for instance, to analyze the results of opinion surveys. Imagine asking 100 people, half of whom are adults, whether they prefer reading books or

How it works

Sunspots

One of the most famous and complicated sets of correlations is between sunspots and a whole range of things on Earth. It is known that sunspots vary regularly over an 11-year period, and this is correlated with a number of economic measures, suicide rates, unemployment levels, the depths of lakes, and many other things. This is almost certainly because of the effect of the Sun's cycle on Earth's weather systems, but the mechanisms are very uncertain.

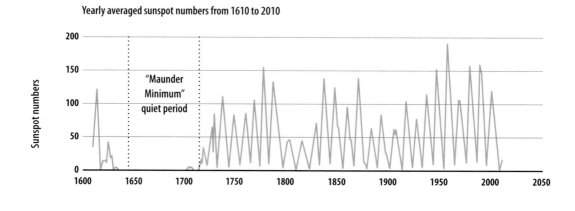

Yearly averaged sunspot numbers from 1610 to 2010

watching TV. This table shows the results:

	Books	TV	Total
Adults	70	30	100
Children	56	44	100
Total	126	74	200

So, both adults and children prefer reading to watching TV (of course). For adults, 70 percent prefer books and for children the figure is 56 percent. But is this a significant difference, or has it arisen by chance? If the figures were more different, the answer would be obvious: if 100 percent of the adults preferred books, but only 51 percent of children did, there would be a clear difference and we could say that adults and children really differ in this way. On the other hand, if 52 percent of adults preferred reading and the figure for children was 51 percent, it would not be safe to draw this conclusion, and it would seem more likely that children and adults think more or less alike about this question. Since the answer is not obvious, we need a statistical test to help us. The best test is

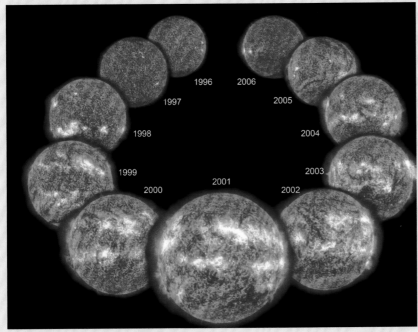

Solar activity, including sunspots, varies on an 11-year cycle.

find out whether our actual values differ significantly from them, so we find the differences between each observed value and the expected one. As we did on page 77, we square the answer, to avoid getting a mixture of negative and positive numbers that cancel out to zero.

the chi- (pronounced "kie") squared test, which was invented by Pearson in 1900. It works out the probability that a particular set of results has arisen by chance.

	Books	TV
Adults	$(70-63)^2 = 49$	$(30-37)^2 = 49$
Children	$(56-63)^2 = 49$	$(44-37)^2 = 49$

Chi-squared in action

Overall, the number of people preferring books is 126 out of 200, which is 63 percent. If there were no difference between adults and children in their views, then we might have expected to see this result:

The figures are all 49 because there are equal numbers of adults and children. Next, we compare these squared differences with the expected values. The simplest way is to divide by the expected value.

	Books	TV
Adults	$49/63 = 0.778$	$49/37 = 1.324$
Children	$49/63 = 0.778$	$49/37 = 1.324$

	Books	TV	Total
Adults	63	37	100
Children	63	37	100
Total	126	74	200

And finally we add up these values:

We call these the "expected" values. We want to

$$0.778 + 1.324 + 0.778 + 1.324 = 4.204$$

Interpreting chi-squared

This number is our chi-squared value, and we need to compare it with the value that would arise by chance. This is a complicated calculation, so usually a set of tables is used. But first, we have to work out the number of degrees of freedom. The number of degrees of freedom is a variable which crops up in many areas of statistics. In our books/TV case, it is found by subtracting 1 from the number of rows (giving 1), subtracting 1 from the number of columns (giving 1 again), and multiplying the answers (giving 1 x 1 = 1). So, we want the first row of the table below.

demonstrate. Let's say you are throwing a dice, and trying to work out the probability of getting each possible score. You start with this assumption:

Score	Probability of getting that score
1	0.167
2	0.167
3	0.167
4	0.167
5	0.167
6	0.167
Total:	≈ 1.000

Degrees of freedom	.995	.99	.975	.95	.9	.1	.05	.025	.01
1	0.00	0.00	0.00	0.00	0.02	2.71	*3.84*	*5.02*	6.63
2	0.01	0.02	0.05	0.10	0.21	4.61	5.99	7.38	9.21
3	0.07	0.11	0.22	0.35	0.58	6.25	7.81	9.35	11.34
4	0.21	0.30	0.48	0.71	1.06	7.78	9.49	11.14	13.28
5	0.41	0.55	0.83	1.15	1.61	9.24	11.07	12.83	15.0

The columns are labeled with the probabilities of our value arising by chance. Our value of 4.204 lies between the two italic numbers, so the probability of our result arising by chance is low: between 0.05 and 0.025 (that is, between 5 percent and 2.5 percent). So, there is probably a real difference between the views of adults and children about book-reading versus TV-watching.

Freedom to change

Degrees of freedom crop up in a lot of statistical tests, and are hard to define but easier to

Notice that the total of the probabilities is 1. That is a certainty—the 100-percent certainty that you will get a score. After some more experiments, you decide that the dice is iffy—it is giving you a lot of low scores. So, using the results of your experiments, you start filling in the table above right with new values, stage by stage. All goes well until stage 6, when putting "0.1" for the probability of getting a 6 gave a total probability of getting some kind of score as 0.90, which is not correct. There is only one possible value that can go in the final box, and that is 0.2, since that brings the total to 1.

Score	Probability of getting the score					
	Stage 1	Stage 2	Stage 3	Stage 4	Stage 5	Stage 6
1	0.3	0.3	0.3	0.3	0.3	0.3
2	?	0.25	0.25	0.25	0.25	0.25
3	?	?	0.15	0.15	0.15	0.15
4	?	?	?	0.1	0.1	0.1
5	?	?	?	?	0.1	0.1
6	?	?	?	?	?	0.1
Total	1.00	1.000	1.000	1.000	1.000	0.90

This shows that although you are free to put whatever probability you think is correct in 5 entries in the lists, you have no freedom at all about what you put in the sixth one. The question mark in red shouldn't really be a question mark at all, it has to be a 0.2 because as soon as you've placed the fifth number, your choice is finished.

Degrees of freedom

This means that the number of degrees of freedom here is one less than the number of data points. But that's not always true. If you take a selection of numbers and find their mean and standard deviation, how much freedom do we have to change those numbers to give the same mean and standard deviation? With five numbers, as in the table on the right, we can change three of them how we like, but then there is just one pair of numbers to complete the list that will give us the same mean and standard deviation. So really we only have 3 choices; 3 degrees of freedom.

	5	7	7
	15	14	14
	56	44	100
	10	9	9
	9	?	3.72
	6	?	11.28
Mean	9	9	9
Standard variation	3.52	3.52	3.52

In this case, the number of degrees of freedom is two less than the number of values. It turns out that however many things we calculate from data (called constraints), we need to subtract that number of things from the number of values to give the degrees of freedom. So, when we calculate mean and standard deviation from a set of numbers, that gives us two constraints, so the number of degrees of freedom = number of values – 2.

SEE ALSO:
▶ Regression, page 118
▶ Tests and Trials, page 160

Skew

ONE OF THE MOST USEFUL APPLICATIONS OF STATISTICS IS THE STUDY OF MONEY. Economists, politicians, historians, and employers are interested in measuring what people earn, how much they save, how prices are increasing, and how debts are mounting up. Because a lot of this data is carefully collected (to work out how much tax to charge people, or to calculate their pensions, for example), it is easy to gather up and analyze.

Often, data related to money follows a normal distribution: the pay per hour to headteachers, postal delivery staff, and firefighters is likely to be normally distributed. Unfortunately, though, there are many exceptions to this. This plot to the right shows how household income in the made-up country of Pretendovia is distributed. There are lots of households with low incomes, and few with high incomes. Whether there are many, few, or a moderate number of households with middle incomes is impossible to say, because,

When lots of people all withdraw money from the bank at the same time, their little fortunes add up to a lot.

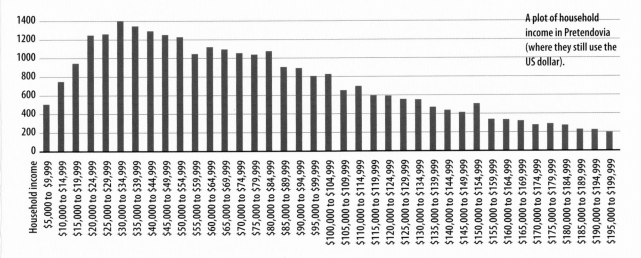

A plot of household income in Pretendovia (where they still use the US dollar).

for distributions like this, "middle" can mean several things. In the military, teaching, sports, and entertainment, there are a few senior or highly skilled people and lots of junior or less skilled ones, and, because these smaller numbers of people are each highly paid, the same shape appears, asymmetrical, with a long "tail" on one side and a peak on the other side. This is a skewed distribution. Skew can be positive, with the tail to the right and the peak to the left, or negative, with tail at left and peak on the right (as shown below).

The average problem

In a skewed distribution, the mean, median, and mode all have different values. For the Pretendovia income data, the mean is $78k, median is $71k and mode is only $33k. This means that, unlike a normal distribution, the "average" can refer to very different numbers. This is particularly important because newspaper headlines and politicians do not always make it clear which average they are referring to, so when someone says "the average pay has increased" or

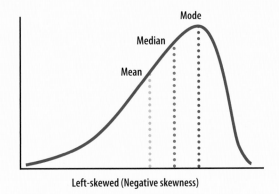

Left-skewed (Negative skewness)

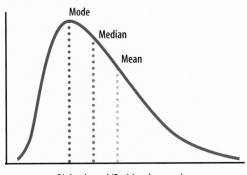

Right-skewed (Positive skewness)

NO EASY ANSWERS

Using the distribution of test grades to judge the difficulty level of a test does raise problems for question-setters and examiners. In a school or a country with an excellent education system which is improved each year, you would expect to see students performing better and better (in fact, if they don't, that would show that the education system is much the same each year). But that would mean that the distribution of results gets more and more positively skewed each year, as each new cohort of students, being better-educated, will find the exams easier than last year's students did. If this situation continued for long enough, it might end up with every student getting an A+, which would make it pointless to set exams at all.

In practice, many education systems set harder exams each year, so the distributions are approximately normal. This raises new problems though. For one thing, since the results will be more or less the same each year, it will not be obvious that there has been any improvement at all. For another, an employer would need to bear in mind how old the qualifications of job candidates are, since someone who got a C last year will probably be better at that subject than another C grader who qualified a few years before, when the exams were easier.

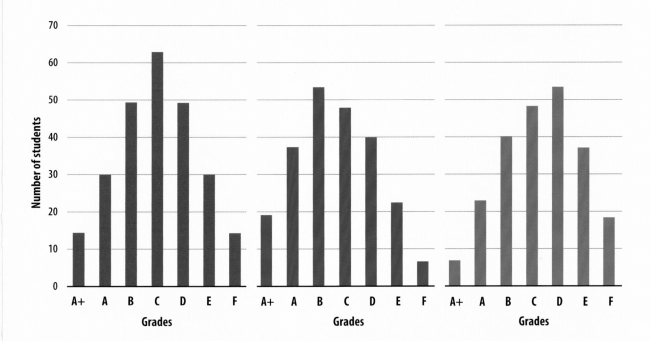

only a few students will get high grades like A+ and A, and a lot will get poor grades; this gives a positively skewed distribution, like the right histogram. A very easy test means a lot of high grades and fewer poor ones, and a negatively skewed distribution, like the middle chart above.

Unskewing

Skewed normal distributions are inconvenient to deal with, but sometimes they can be converted to more normal versions. However, just as when dealing with outliers (see page 108), tampering with data should only be done after looking at where that data comes from. Is there some reason to think that the population itself is skewed? If it is biological or financial data, then this is sometimes just what to expect. For instance,

many plants, like some varieties of grapes and strawberries, produce fruit with a range of sizes (much to the annoyance of supermarkets, who like standard-sized fruits because they are easier to price and package). If they do, then the distribution of these sizes is usually strongly positively skewed, with just a few large fruits and lots of small-to-medium ones. Surprisingly, a simple mathematical formula can predict this skew, and, even more surprisingly, the same formula works for economic data, too, and also for writing: the lengths of postings on social media follow the same pattern.

Lognormal

The left-hand histogram overleaf shows the recovery times of 1,000 patients from the same

Non-Parametric Statistics

A GREAT DEAL OF STATISTICS INVOLVES WORKING THINGS OUT ABOUT A POPULATION FROM A SAMPLE OF DATA TAKEN FROM IT. The population might be an actual human population, a cloud of molecules, or a database of medical reports.

Often, the characteristics of the population are distributed according to the normal distribution. This is true, for example, of the speeds of all the molecules in a cloud. In these cases, statistics can be simple enough. If we take a large (and unbiased) sample of the molecules in a cloud, and measure all their speeds, then we can safely assume that the values of statistics of our sample, like the mean and standard deviation of the speed, are close to the mean and standard deviation of the whole population. These population values are called "parameters," from Greek words meaning "almost measurements."

Distributions unlimited

However, many distributions are not normal. Human height is not, because men tend to be taller than women. But this is an easy problem to solve: we can simply

Charles Spearman, who developed a powerful test for correlation.

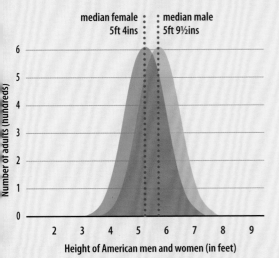

median female
5ft 4ins

median male
5ft 9½ins

Number of adults (hundreds)

6
5
4
3
2
1
0

2   3   4   5   6   7   8   9

Height of American men and women (in feet)

calculate separate distributions for men and women, and these will be normal, as shown in the chart above. Some things are distributed according to distributions that are known, but are not normal ones. Examples are the binomial and Poisson distributions (see more on pages 38 and 100). These are well understood and, in many circumstances, are very similar to normal distributions. Other distributions are very different to normal ones, like the ones shown on the right.

Mystery populations

Often, we just don't know how the population is distributed. Several tests have been developed to use where either the population is not normal, or is unknown. Because these tests do not rely on or assume any of the parameters of the population, they are known as

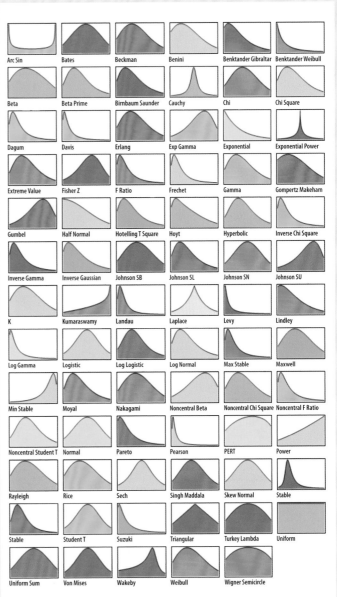

The many shapes of
data distributions.

www.wolfram.com/mathematica/new-in-8/parametric-probability-distributions/univariate-continuous-distributions.html

non-parametric tests. One quite simple and powerful non-parametric test for correlation (see page 122) is Spearman's test or Spearman's rank correlation. (The word *rank* comes from an early 15th-century French word for a row of soldiers.)

Spearman in action

Let's say we want to find out whether students who are good at math are also good at physics, based on their exam marks. As we have seen in the box on page 136, we can't assume that such marks are normally distributed; they might be very skewed. Spearman's test overcomes this problem by using just one fact about each item of data—its rank. If the data we have is as seen in the second and fourth columns below, then they are ranked as in the third and fifth columns. Once the data has been ranked, the math rank is subtracted from the physics one. Just as in

the case of least-squares fitting, each difference is squared to avoid the problem of negative differences canceling out positive ones (see more about that on page 76).

The squared differences are then added up:

$$1+1+0+4+0+0+0+4 = 10$$

Now we use Spearman's formula:

$$p = 1 - \frac{6\sum d_i^2}{n(n^2-1)}$$

Here, $\sum d_i^2$ is the sum of the squared differences (**10**), and n is the number of pairs of data (**8**), so:

$$p = 1 - \frac{6 \times 10}{8 \times (8^2-1)} \approx 0.88$$

Student	Math score	Math rank	Physics score	Physics rank	Difference in ranks	Square of difference
George	88	4	69	5	−1	1
Bill	87	5	70	4	1	1
Keiko	77	6	66	6	0	0
Dana	98	1	77	3	−2	4
Alex	52	7	51	7	0	0
Jan	94	2	81	2	0	0
Omar	22	8	10	8	0	0
Hilary	92	3	82	1	2	4

Since Spearman's test can be used for any kind of data, and is very simple, why does anyone use the Pearson coefficient? The reason is that, in simplifying the data, we discard a lot of information. For instance, in the table on page 142, the scores for George and Bill are very close, and it may be that a repeat of this test with a new set of exam results might reverse their ranks. Meanwhile, Alex and Omar are much worse than the others at both subjects, so perhaps they should be excluded. The question is whether skill at math is related to skill in physics, not whether people who do poorly in one subject are bad at the other. Although the latter is a possible conclusion, this analysis is not designed to show it. That question could be investigated with more complicated statistical tools. Another advantage with Spearman's approach is that it can handle data which is not in the form of numbers. In a subject like math it is easy to use percentages to score results,

How good are these musicians? Statistics will help answer that question.

since the person who grades the papers can simply count how many questions are answered correctly. So, a B grade might mean "between 70 and 85 percent questions correct." But in a subject like art, dance, or music, where the quality of a performance or composition must be judged, there may be no numbers to use.

As with the Pearson coefficient, a value of 0.88 indicates a strong correlation between skill in math and in physics. The students who are good at one, are good at the other. (See more, box above.)

Trouble with ties

One problem with letter-based grades is that there aren't many of them, so different people often get equal ranks, or tie. The solution is shown in the table overleaf. The italicized ranks are averages due to two students being tied and getting the same ranks in a subject. Spearman's formula gives us:

$$p = 1 - \frac{6 \times 27.5}{9 \times (9^2 - 1)} \approx 0.77 \text{ – a strong correlation.}$$

The intelligence question

Charles Spearman had been a captain in the British army until his love of science

Student	Music grade	Music rank	Adjusted rank	Math score	Math rank	Adjusted rank	Difference	Squared difference
Anita	E	8	8	58 percent	9	8	0	0
Bob	A+	2	2	88 percent	3	3	−1	1
Carol	A	3	3	90 percent	2	2	1	1
Dave	F	9	9	50 percent	9	9	0	0
Jamila	D	7	7	70 percent	5 or 6	5.5	1.5	2.25
Fergus	C	6	6	66 percent	7	7	−1	1
Gail	B	4 or 5	4.5	92 percent	1	1	3.5	12.25
Femi	B	4 or 5	4.5	70 percent	5 or 6	5.5	−1	1
Isobel	A++	1	1	77 percent	4	4	−3	9
						Sum		27.5

led him to resign in 1897 to take a PhD in psychology. Like many other great statisticians, his breakthroughs in statistics came about as a result of his determination to solve the mysteries of another area of science. In Spearman's case, the topic which fascinated him was the nature of human intelligence. Spearman and others of his time wondered whether it really makes sense to talk of human intelligence at all. People have many different mental skills, from breaking codes to doing crosswords and from spotting differences between similar images to working out riddles, but is there any link between these skills that could be described by the general term "intelligence?" The evidence at the time said that the answer was no. Statistical analysis of the performance of people on different tests showed that there was little correlation between the results. Instead of carrying out more tests or designing new ones, Spearman looked at the statistical methods used to analyze the tests, and found that they were flawed. He devised a method (called attenuation) to improve them, and discovered that, in fact, performance on different tests is correlated. So, there is something that links them all together.

Clash of the titans

Although he was a great psychologist, Spearman was not very tactful, and he was very critical of the methods of Karl Pearson (who was very easily upset). So when, in 1904, Spearman wrote a paper in which he said that Pearson's methods were both overcomplicated and incorrect,

DOUBLE INTELLIGENCE

Spearman pioneered a new theory of human intelligence which is still largely accepted today. His idea is called the two-factor theory, and it claims that each of us has two kinds of intelligence: G (for "general") intelligence and S (for "specific"). Working out logic puzzles requires a kind of S intelligence that uses verbal skills; mental arithmetic skill is another kind of S intelligence, and seeing how to get through a maze is a third. In addition to our particular supply of S intelligences, we each have some level of G intelligence, which forms a foundation of the S skills and is used when tackling all kinds of challenges.

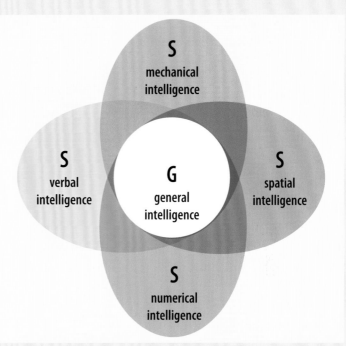

Pearson was decidedly outraged. Pearson criticized Spearman in his turn, and in particular he claimed that Spearman's corrections to the statistical test of intelligence were wrong. The two men never forgave each other and continued to argue in print for the rest of their lives.

Error correction

Actually, even if they had been the most polite and charming scientists on Earth, Spearman and Pearson were almost certain to have fallen out. Central to all of Pearson's work was the—incorrect—belief that all populations were distributed either according to the normal distribution or to distorted versions of it, and that the only four statistical parameters (called the four moments) that were ever needed were mean, standard deviation, skew, and a fourth measure, called kurtosis, which defines the spread of the lower part of a distribution. He thought that if these four parameters are known, everything useful about a population can be worked out from a large sample. Spearman's rank correlation test showed that none of the four parameters are essential. Meanwhile, other statisticians showed that, even worse for Pearson, the four parameters could never be known exactly in any case, unless the entire population was analyzed. Sampling a few—or even many—would not do.

SEE ALSO:
▶ The Shapes of Data, page 38
▶ The Average Human, page 96
▶ Skew, page 132

Compare and
Contrast

TREASURY OF HUMAN INHERITANCE. INSANITY AND ALLIED CHARACTERS. PLATE XIX.

Issued by the Francis Galton Laboratory for National Eugenics.

ONE QUESTION WHICH COMES UP AGAIN AND AGAIN IN STATISTICS, and in many other aspects of life, too, is: "Is this the same as that?" Are the cherries the market trader gives you as good as the ones on display? Is that "10,000-hour light bulb" really what it says on the box? Is this coin, or are those dice, fair?

Tackling these questions is a job for a statistician, and, if that statistician was Karl Pearson (see more, page 122), finding the answers was either easy, or impossible. Pearson was firmly of the

Karl Pearson at work below left, and the fruits of his work above left, a page from *Treasury of Human Inheritance* of 1912.

The Guinness Brewery in Dublin, Ireland, at the turn of the 20th century, and the plaque still there in honor of William Sealy Gosset, the brains behind the Student t-test (see overleaf).

opinion that every kind of data worth considering was distributed according to the normal distribution or distorted versions of it, which could be completely described by a handful of numbers, of which the mean and standard deviation (see page 68) were most important. All that was required was a large enough sample of the data; typically at least 30 values. However, if the sample was smaller than this, then questions could not be answered.

Small problems

For a young university graduate called William Sealy Gosset, this was a problem. He had been working for the Guinness beer brewery in Dublin since 1899 as a statistician, a member of a team whose task it was to find a reliable, scientific way to identify which of the many hops that Guinness used made the best-tasting beer. Gosset's main job was to make sure that different batches of hop mixture, from different parts of the country, were similar enough to ensure that the beer made from them would taste the same. His difficulty was that all he had to go on were about a dozen sample measurements from each batch.

Rough workings

In those days, Pearson was the King of Statistics, and he ruled his mathematical empire from his Statistical Laboratory at University College London. Gosset knew all about Pearson's views on small samples, but he also knew that, even with a few numbers in a sample, the mean and standard deviation of the population could be used as a rough estimate. Gosset set out to work out how rough that estimate would be (by calculating a quantity called the standard error of the mean). He made some progress and then, rather bravely, he obtained permission from his boss to go to Pearson for help.

A new distribution

Luckily, Pearson was impressed with Gosset, and helped him work out the effect of sample size on the accuracy with which the mean can be estimated (for instance, how accurately one could estimate the mean weight of the sheep in a flock if one only knows the weights of six of the sheep). The spread of estimates of the mean forms a pattern called the t distribution. The shape of this distribution depends on the size of the sample and, for large samples, the t distribution becomes the normal distribution.

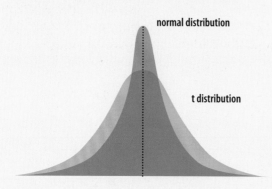

normal distribution

t distribution

The t distribution for a sample size of 12, shown here, is broader than a normal distribution. From his work, Gosset developed a new kind of test; the t-test, which compares the mean values for two groups to discover whether they are significantly different, and works whether the groups are small or large.

The t-test in action

The t-test works like this. After each playing games with several other football teams, Ravens and Eagles have the following scores. Can we say whether Ravens or Eagles is the better team?

Scores

	Ravens	Eagles
	1	2
	3	2
	3	1
	0	1
	2	2
	3	
Mean	2	1.6

To find out, we first find the difference between each score and the mean, and then square that value. So, for the Ravens' first game, this squared difference is $(1-2)^2 = 1$. Calculating this value for all the scores gives this table of squared differences, which also adds up all the squared differences for each team:

Squares of differences

	Ravens	Eagles
	1	0.16
	1	0.16
	1	0.36
	4	0.36
	0	0.16
	1	
Total	8	1.2

These totals are called variances; (variance is the square of the standard deviation). We now calculate a statistic called the pooled sample variance. It combines the variances of the two samples, to give a single number, and is sometimes written S_p^2

By degrees

First we must work out the number of degrees of freedom (see more, page 130). We are just calculating one statistic from the data for each team, so the number of degrees of freedom for each team is just 1 less than the number of results for the team, so that is $6 - 1 = 5$ for the Ravens and $5 - 1 = 4$ for the Eagles. We add these to find the total number of degrees of freedom, which is 9.

$$S_p^2 = \frac{v(\text{Ravens}) + v(\text{Eagles})}{\text{number of degrees of freedom}} = \frac{8 + 1.2}{9} \approx 1.022$$

(v = variance)

Now, we use the formula for t:

$$t = \frac{\text{mean of Ravens' scores} - \text{mean of Eagles' scores}}{\sqrt{S_p^2 \times \left(\left(\dfrac{1}{\text{no of Ravens' games}}\right) + \left(\dfrac{1}{\text{no of Eagles' games}}\right)\right)}} \approx \frac{2 - 1.6}{\sqrt{1.022 \times \left(\left(\frac{1}{6}\right) + \left(\frac{1}{5}\right)\right)}} \approx 0.653$$

To work out what this t-value means, the simplest approach is to refer to a table.

		Probability		
		90%	95%	99%
	1	6.314	12.706	63.657
	2	2.92	4.303	9.925
	3	2.353	3.182	5.841
Number	4	2.132	2.776	4.604
of degrees	5	2.015	2.571	4.032
of freedom	6	1.943	2.447	3.707
	7	1.895	2.365	3.499
	8	1.86	2.306	3.355
	9	*1.833*	*2.262*	*3.25*
	10	1.812	2.228	3.169

The italicized 9 degrees of freedom row is the one we are interested in, and it shows that, for instance, given a t-value of 1.833, there is a 90 percent chance that the two teams are significantly different. Since our t-value is much lower than 1.833 we can be confident that the teams are no different.

Go Ravens! Go Eagles! The numbers show it's still all to play for.

Ronald A. Fisher looks thoughtful, above, as do the students being tested, below.

(though this word also refers to different kinds of tests). Gosset's discovery was a great personal triumph, too, so, naturally enough, he wanted to make it public. But his managers at Guinness were not so keen. For them, the new tests were a secret weapon and they had no wish to let their competitors in on it. But they were not unreasonable, and so they allowed Gosset to publish, but not to give his name on the report. So Gosset called himself Student and his test is usually called the Student t-test today.

Fisher of facts

In fact, neither any other brewers, nor anyone else, seemed to have noticed Gosset's breakthrough, except for another young statistician called R. A. Fisher. Much as Galton's work inspired and was greatly improved by Pearson, so was the case as Fisher took up Gosset's ideas and developed them further. Gosset remained on friendly terms with both Pearson and Fisher—no one else seemed

Fake student

As well as being of great practical use, Gosset's discoveries were important for another reason. For the first time, a test had been developed that did not refer to some other population; it was entirely stand-alone. This type of test is sometimes called an intercomparison

OTHER Ts

The full name of the test discussed here is the independent samples t-test; there are two other kinds, also developed by Gosset. The paired sample t-test compares means from the same group at different times, to test whether the group has changed significantly. And the one sample t-test compares the mean of a group against a known mean, for instance, checking whether the mean age of a particular pride of lions is significantly different from the world average.

able to manage this tricky feat. This might be partly due to his humble attitude to them both. He used to say to admiring fans of his work (of whom there were many), that "Fisher would have discovered it all anyway."

ANOVA

Among other things (see more, page 160), Fisher extended the t-test so that many more means could be compared, and in so doing he developed a technique called analysis of variance (abbreviated to ANOVA) which is one of the most powerful of all statistical techniques. Let's say a research team is trying to find out whether diet has any effect on math test scores. They want to try the effects of vitamin C, coffee, and omega-3 oil. They test these on 15 students, giving extra amounts of one of the three trial substances to five students each, and scoring the tests out of 100 (in reality many more students would be used, but the idea is easier to see with just a few). Maybe the results look like this:

Scores of students with extra
Vitamin C: 75, 76, 88, 92, 100
Coffee: 74, 79, 80, 98, 99
Omega-3: 59, 78, 79, 88, 98

In this case, there isn't much difference between the groups, but quite a wide range within the groups. Without going into details here, if we calculated the variance within each group it would be quite large, but the variance between the groups would be quite small.

A second attempt

Now the research team tries again. They record the amount of sleep each student had the night before, and classify them that way. Now the results look like this:

Scores of students with
under 5 hours sleep: 48, 52, 59, 68, 72
5 to 8 hours sleep: 79, 88, 94, 97, 100
more than 8 hours sleep: 66, 74, 79, 81, 86

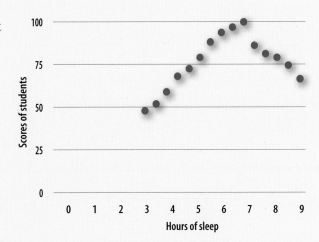

How it works

Jackknife

The word "statistic" refers to a value, such as the mean, standard deviation or skewness, calculated from a sample of data. But what we really want is information about the population from which the sample is drawn. For instance, if we are interested in the weights of ostriches, we might weigh a hundred of them (the sample), average the values and hope that the result is close to the mean weight of the world's ostrich population. That is, we want the mean of the sample to be close to the mean of the population. The difference between these means is called the bias. The jackknife is a technique we can use to estimate bias. We first calculate the statistic for the whole population, then subtract one data point from the population and calculate the statistic again. We repeat this until all the data points have been removed. Finally, we average the results of all these calculations. This average is then used to estimate the bias. Like a real jackknife (or penknife), this technique can be applied to many tasks. It usually needs at least 1,000 data points to work properly.

P(it's too heavy to fly) = 1

The differences between scores within groups are quite small, but the differences between them are large. This suggests that the amount of sleep does affect performance. In this case, only ANOVA could have found this result; calculating the correlation (see more, page 122) would not have revealed the link because the benefit of sleep does not increase steadily with amount of sleep. The detailed calculations are quite fiddly, but ANOVA boils down to calculating this sum:

$$\frac{\text{variance between groups}}{\text{variance within groups}}$$

The larger the result of this is, the more likely there is a real effect.

Computer world

Although ANOVA is a very powerful test, it involves so many calculations that computers are needed to make full use of it. Today, that is easy, but the arrival of cheap and powerful processing power has also allowed the development of many other tools. Some of these tools are based closely on the approaches of Gosset and Fisher, but they also use the idea of simulation. The Monte Carlo test (see more, page 58) was an early example of simulation. These are known as re-sampling tests because they return to the same data many times. Tests like these are very useful where we have a great deal of data but it's not clear how relevant some of it is. Environmental modeling and studies of ecosystems are of this type.

THE BOOTSTRAP

The bootstrap (which is probably the most-used resampling technique today) is similar to the jackknife, but rather than finding improved values of statistics, it is used to calculate how good they are as they stand. Again, it recalculates statistics many times, ignoring one data point at a time. It ends up with confidence limits. If these are inadequate, it tells us how big a sample we need.

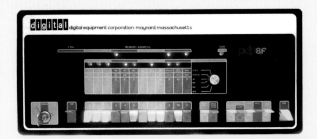

Left: A PDP-8 microcomputer, which became popular in the late 1960s, and which was often used for statistical analysis.

A GIS, or "geographic information system," is built from a wealth of statistical data about landscape, weather, and other environmental phenomena.

SEE ALSO:
▸ Measures of Spread, page 68
▸ Correlation, page 122
▸ Tests and Trials, page 160

Measuring Confidence

EARLY STATISTICIANS WOULD HAVE BEEN AMAZED HAD THEY KNOWN JUST HOW MUCH THEIR SUBJECT WOULD GROW, and the changes it would make to the world. For instance, it has made many words and ideas much more meaningful and useful by showing how we can express them in terms of numbers. One of the most widely used terms in statistics is "confidence."

A lovely bunch of coconuts.

Let's say a team of botanists wants to find the mean weight of the coconuts on an island. Weighing just one coconut won't really tell them much—they might randomly pick a large or small one. Weighing them all will give them the exact answer but is too big a job. So, they decide to weigh 30 coconuts, and find the mean weight is 20.5 ounces. This weight won't be exactly the same as the mean weight of all the coconuts on the island. If they weighed another batch of 30 and found a mean weight of 20.6 ounces, it wouldn't surprise them at all.

Jerzy Neyman was the man who gave us all confidence.

Margin of error

So, rather than giving their value of the mean weight as a single number, it would be better to give it as a range; something like "the average weight of the coconuts on this island is between 18 and 22 ounces" perhaps, which is more briefly written "20 ± 2 ounces." The ± 2 is known as the margin of error.

Estimating averages

To work out the actual values here, the team first needs to know how much the coconuts on the island differ in weight. If they are all quite similar in weight they should be able to get quite a good estimate of the average weight. To get a handle on the spread of weights, they need the standard deviation of the coconuts on the island. However, they don't know this, so they have to use the standard deviation of the sample they have, which is 5.1 ounces. Now all they need is a formula for the margin of error, and, luckily, Polish mathematician Jerzy Neyman worked one out in 1925:

$$\pm \frac{Z \times s}{\sqrt{n}}$$

(**s** is standard deviation and **n** is sample size). Say the value of **Z** is 1.96 (we'll see why later), this becomes

$$\pm \frac{1.96 \times 5.1}{\sqrt{30}}$$

which is ±1.8, so the biologists conclude that the mean weight of the coconuts on the island is 20.6 ± 1.8 ounces. This does not mean, however, that the mean weight is absolutely certainly in this range. This is where the Z comes in. With Z set to 1.96, the biologists can say that they are 95 percent confident that the mean weight is 20.6 ±1.8. Z-scores are explained on page 161.

NOT NORMAL

The biologists don't really know whether the coconut sizes on the island are distributed normally. However, thanks to the central limit theorem, (see page 51), we know that the means of samples will still be distributed normally even for data which is not normal, as long as we have a fairly large (at least 30) dataset.

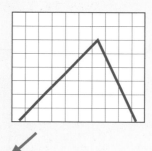

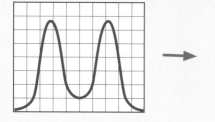

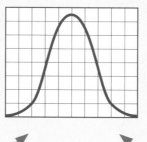

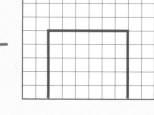

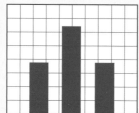

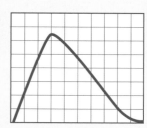

How confident?

What does "95 percent confident" mean? That, if the biologists carried out their sampling study 100 times and therefore got 100 sample means, they would expect 95 of those means to lie in the range 20.6 ± 1.8. More briefly, they would say that 20.6 ± 1.8 is the 95 percent confidence interval. Of course, 95 percent confidence is a long way from certainty. Can the biologists improve their answer? There is an easy way to do this, and a harder way. The easy way is to change

the value of Z. For a confidence level of 99 percent, for instance, the Z-value is 2.576. Putting this in the formula gives a margin of error of:

$$\pm \frac{2.576 \times 5.1}{\sqrt{30}}$$

which is ± 2.4, so the biologists can be 99 percent confident that the mean weight of the coconuts on the island is 20.6 ± 2.4 ounces. This kind of answer may not please the biologists though; they have paid for their extra confidence by increasing the range, which is a bit like changing from "I might possibly see you on Tuesday" to saying "I'll almost certainly see you on Monday, Tuesday, or Wednesday." The less easy way to improve the answer is to carry out a larger survey. If the biologists measure 100 coconuts instead of 30, and get a mean of 20.1 and a standard deviation of 5, the 95 percent confidence interval becomes

$$20.1 \pm \frac{1.96 \times 5}{\sqrt{100}}$$, which is 20.1 ± 0.98 ounce.

Karl Pearson, second left, has a hand in the story of confidence, of course, but so does his son Egon, pictured here at the age of two sitting on his pop's lap. Big sister Sigrid, far right, does not look very impressed. She had no interest in statistics and became a poet instead.

Jerzy Neyman

Confidence in statistics was first defined in 1925 by Polish mathematician Jerzy Neyman. Like Siméon Poisson, Neyman loved physics but, also like Poisson, he was too clumsy to be good at the experimental side of the subject. He came across a book called *The Grammar of Science* by Karl Pearson (a volume that also inspired Albert Einstein) and decided to remain at Kharkov University in Ukraine to study mathematics in depth. However, he fell ill and went to the Crimea to recover in 1920. There he met Olga Solodovnikova. They returned to Kharkov and married, but Poland and Russia were now at war and, a few days later, he was imprisoned for six weeks. Although he returned to the university afterward, the political situation was so uncertain

How it works

Error bars

Error bars can be used to indicate the accuracy of a measurement (as on page 94). But often, rather than individual measurements, a set of mean measurements will be available, such as the mean speed of cars on a road. Perhaps there is a mean value available for each day of the week.

Looking at the data alone, it's clear that the speed is highest on Saturday and lowest on Sunday, but we need more than this to say whether these differences are significant. Adding error bars, whose lengths are 95 percent confidence limits, allows us to tell this at a glance.

So, it looks like the higher mean speed on a Saturday is only a random variation, but we can be confident that people drive significantly more slowly on average on a Sunday. Error bars do not always show confidence limits; sometimes standard deviations are used instead, but confidence limit bars tell us more, especially when the number of values used to calculate the means is small.

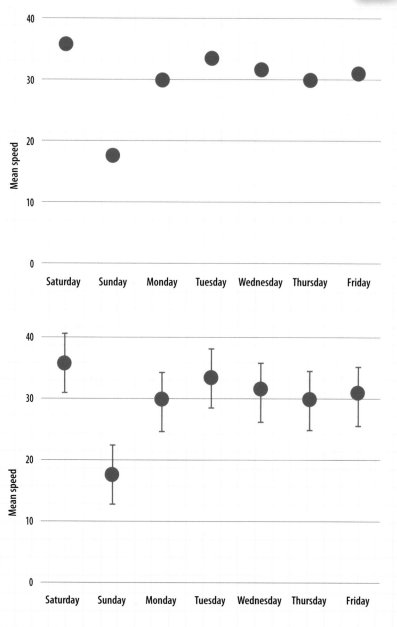

that he was always at risk of arrest, and he had to go into hiding at least once until it was safe to return. Finally, he found a safer place to work, at the College of Agriculture in Warsaw. He also worked for the weather forecasting service there and was able to focus his studies on statistics.

Meeting heroes

In 1925 Neyman was awarded the funds to work with Pearson at his London statistical laboratory. Although Pearson was a statistical hero of Neyman's, he was now 68 and out of touch with modern mathematics. However, Pearson's son Egon was also a mathematician, and he and Neyman worked together. Inspired by the work, Neyman returned to Poland and set up a statistical laboratory in Warsaw, until the approaching World War II

FAILURE

When a new process is introduced in a factory, it is rarely possible to make it completely reliable. There will be some misshapen chocolates, broken eggs, wonky bolts, or bottles with no labels. In such cases, the most useful way to express the confidence level in the process is in terms of likely number of failures, as this table shows. So, for instance, a 95 percent confidence limit means that a failure rate of 1 in 20 is to be expected.

Confidence limit	90	95	99	99.9	99.99
Risk: 1 failure in n events	10	20	100	1,000	10,000

meant he had to flee again—this time to the United States, where he set up another new statistical unit and remained for the rest of his life.

Risk

A major reason for using confidence limits is to make clear the risk involved in making the a decision. Putting a number on risk is simple once confidence limits are known: it is defined as Risk = 100 percent minus confidence level. So, a 95 percent confidence level corresponds to a 5 percent risk, and to reduce risk to 1 percent, a 99 percent confidence level must be achieved.

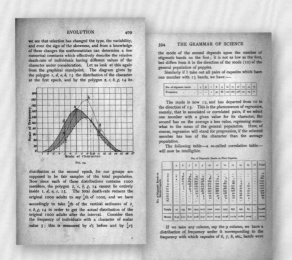

Karl Pearson used statistical techniques a lot in his 1892 book *The Grammar of Science*.

SEE ALSO:
▶ Least Squares, page 74
▶ Outliers, page 108

Tests and Trials

Ronald Aylmer Fisher, seen here as a young man, developed most statistical tests. He is usually referred to as "R. A. Fisher," since that is how he gave his name on his publications. It's possible that he was slightly embarrassed by his odd middle name, Aylmer (his first name is the very respectable Ronald). Fisher's parents had seven children; the first were called Geoffrey and Evelyn and the third was Alan. When Alan died, Mrs Fisher decided that all their children from that time on would have a "y" in their names.

IMAGINE A NATURALIST WANTS TO FIND OUT WHETHER THE NUMBER OF GIANT SPIDERS IN HOUSES HAS INCREASED SINCE LAST YEAR, when the mean number of giant spiders per house was 90 (this statistic is called the "population mean"). The naturalist randomly selects 40 houses, and finds that the mean number of giant spiders is 110 (this is called the "sample mean"). The question is, what does this increase mean? Would we expect it to happen by chance, or does it mean more than that?

It's not enough to know that there are more spiders in this year's sample mean than last year's population mean, since we would not expect to

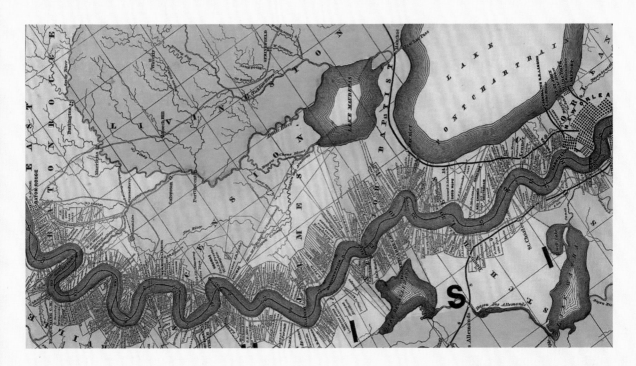

Z-scores can be used to show whether a river is unusual in its shape, length, or amount of pollution.

get exactly the same mean for our sample as there is in the population. (The difference is known as the "sample error.") If we repeated the experiment with a different 40 houses, the second sample mean might be less than 90. So, how do we find out if there really is a giant spider invasion?

Z-scores

One way would be to take many samples of the population, note the mean of each sample, and plot all the means. But there's a much quicker way: we can look at last year's data, and see whether our new sample fits into that pattern. In other words, we find the probability that if we

Don't panic! Statistics will save us from an invasion of spiders.

randomly select one of the values from last year's data, that value would be 110. To do this we first need a statistic called the Z-score. This simply measures distances from standard deviation, in the units of that deviation. Switch scenarios from spider invasions to the course of rivers: if the mean length of a collection of rivers is 30 miles, and the standard deviation is 10 miles, then a river 20 miles long is 1 standard deviation from the mean, and the Z-score of that river is 1. Z is defined as:

$$Z = \frac{\textit{Sample mean} - \textit{population mean}}{\textit{Sample standard deviation}}$$

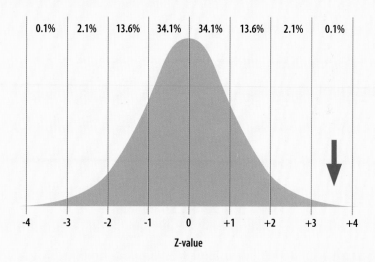

Z-scores along the base, that shows the position of the 3.61 Z-score calculated. The percentages along the top show how much of the whole area of the distribution is contained in each stripe.

Going back to our spider investigations, we already know the top part of this as:

110 – 90 = 20.

Sample standard deviation

The sample standard deviation can be estimated as:

$$\text{sample standard deviation} = \frac{\text{population standard deviation}}{\sqrt{\text{sample size}}}$$

So,

$$Z = \frac{\text{Sample mean} - \text{population mean}}{\text{population standard deviation} / \sqrt{\text{sample size}}} = \frac{110 - 90}{35 / \sqrt{40}} = 3.61$$

From Z-score to probability

So, we know that our sample value is 3.61 standard deviations away from the population mean, but we need to relate this to a probability. The chart above shows a normal distribution, marked with

Under the curve

By way of illustration, if the distribution represented the heights of 100,000 men, then we would expect 50,000 of their heights to appear in the left half of the diagram, 100 height measurements in the first stripe on the left, 2,100 in the second, 13,600 in the next, and so on, with 100 heights located in the final stripe on the right. This also means that if we pick one of those 100,000 men at random, there is a 0.1 percent chance that his height will appear in the final stripe.

Spider chances

Back to our spider survey, since we are dealing here with last year's spider populations, that means that, if the houses in our sample had an average of 110 spiders, then last year there would have been a 0.1 percent chance that the sample would appear in that final stripe of the distributed data. Since there is only a 0.1 percent chance of it appearing there by chance, we can conclude that there is a

POPULATIONS VERSUS SAMPLES

The statistics that describe a population, such as its mean or standard deviation, are called parameters. Usually, these are unknown, and we have to work with the statistics of a sample instead, and as long as we have a large enough sample (usually 30 items is enough), we can do this. The population mean is simple. Thanks to the central limit theorem (see page 51) we can assume it is about the same as the sample mean. For standard deviation, things are not quite so easy, because the smaller sample, the more variable the standard deviation becomes.

This is why the sample size appears in the equation

$$sample\ standard\ deviation = \frac{population\ standard\ deviation}{\sqrt{sample\ size}}$$

The reason that the square root of the sample size is used is that, as sample size increases, each new increase makes less difference to the statistics. For example, these four histograms show the distributions of 30, 60, 90, and 120 randomly selected values. The first two patterns look quite different, but the last two are more similar.

Also, if we compare the square roots of the numbers in the table below, we can see that the last two are more similar than the first two.

Number	Difference of number	Square root (approx)	Difference
30		5.48	
60	30	7.75	2.27
90	30	9.49	1.74
120	30	10.95	1.46

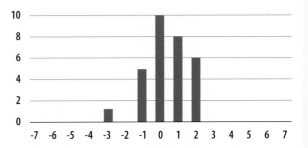

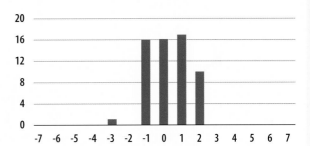

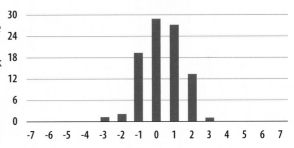

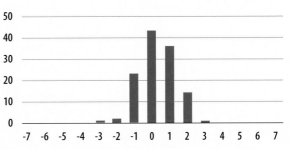

real and meaningful increase in the number of spiders in the neighborhood.

Significance of significance

If a result is said to be "significant," that usually means "at the 5 percent significance level." So far,

we have been interested in a significant increase in giant spiders, asking the question: "Is the number of spiders in our sample in the top 5 percent of a normal distribution?" This top 5 percent is shown in blue in the topmost chart below left.

A different question

But, we could have asked a different question: "Has there been a significant change in spider numbers?" This is two questions in one: "Are there significantly more spiders now?" and "Are there significantly fewer spiders now?" So, now there are two blue areas of interest, shown in blue in the lower chart at left. If our sample is in the right area then there are significantly more spiders, if it is in the left one there are significantly fewer. This diagram, with its pair of blue "tails," shows why this is called a two-tailed test, as opposed to the one-tailed test we began with.

How many tails?

A key point to make is that the total area of the two blue tails is 5 percent, so each is now 2.5 percent, half the size of the single tail that corresponds to our original question. Imagine our sample had a Z-score of 1.8. Does this suggest a significant increase? Yes, because 1.8 lies within the blue area on the single-tail chart. But we could ask

The same data can be tested with one or two tails. The answer depends on the question.

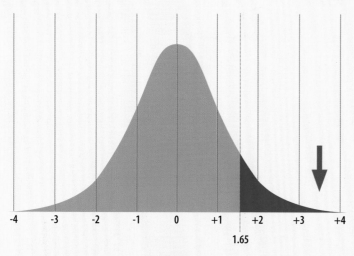

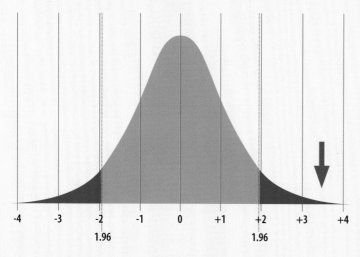

SIGNIFICANCE

Very often people ask whether something is statistically significant. Significance level is 1 minus confidence level, so a confidence level of 95 percent (0.95) corresponds to a significance level of 5 percent (0.05). Significance level is also sometimes called alpha, or alpha level. When someone says that something is statistically significant, that usually means its confidence level is greater than 95 percent. However, there's nothing special about 95 percent, and sometimes other levels are used, so to be clear, the alpha level should always be stated.

again: "Does it suggest a significant change?" and answer no, because it does not lie within either of the blue tails on the second chart.

Which question and answer?

This seems very odd, but it is the price we pay for seeking more information from our data. The broader the question, the more evidence is needed to support it. This shows how important it is to be precise in the questions we ask in statistics. An obvious solution would seem to be to use both tests on the same data, but it turns out we cannot. We need to decide in advance of carrying out the analysis which question to ask, and we cannot ask both, for a reason to be explained on page 168.

The taste of tea

The statistics of trials and experiments are largely the work of Ronald Aylmer Fisher. Although Fisher studied mathematics and astronomy at Cambridge University, he was also very interested in biology, and became even keener after working at a farm for a while following his degree course.

The taste of tea drove an English mathematician to develop statistical tests. What else would do it?

which hosted long-running studies of the effects of soil chemistry, fertilizers, and crop strains on harvest yields and livestock. It was there that he developed a whole range of tests for use in scientific research. To explain how some of his tests worked, in 1935 Fisher told a story of a woman of his acquaintance who claimed that tea tastes different depending whether milk is added to the cup before or after the tea. Fisher showed

Fisher first got a job as a statistician at a London investment company, but he left in 1919 when the great statistician Karl Pearson offered him a job at the Rothamsted Experimental Station,

Below: The bearded figure of R. A. Fisher, third right, is unmistakable at this outdoor tea party at Rothamsted. The woman facing him may be Dr Muriel Bristol, the subject of his experiment.

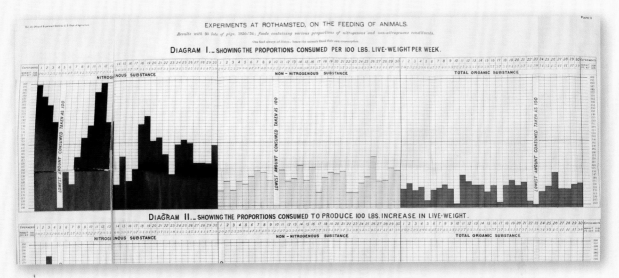

EXPERIMENTS AT ROTHAMSTED, ON THE FEEDING OF ANIMALS.

DIAGRAM I.— SHOWING THE PROPORTIONS CONSUMED PER 100 LBS. LIVE-WEIGHT PER WEEK.

DIAGRAM II.— SHOWING THE PROPORTIONS CONSUMED TO PRODUCE 100 LBS. INCREASE IN LIVE-WEIGHT.

A chart representing data on the effect of food supplements
on the growth of pigs as studied at Rothamsted in the 1850s.

how to calculate the number of times the woman would have to judge correctly to prove her claim. Although this sounds unlikely, this really happened, and it may have encouraged Fisher's interest in statistical testing. Perhaps because he was jealous of Fisher's success, Pearson soon grew to dislike him. In 1917, Pearson had criticized a paper Fisher had written two years earlier and Fisher took offence. Actually, Fisher's papers were very often criticized, and one reason is that they were the first to apply statistical tools to problems in biology. Fisher believed that the problem was that statisticians, though often excellent physicists, knew little of biology, while biologists were often poor mathematicians. Thanks to Fisher, this has completely changed today, and biology makes far more use of statistics than does physics, chemistry, or any other science. In 1922 Pearson definitely went too far, claiming that Fisher had damaged the reputation of the whole of statistics by publishing papers containing errors. This did real damage to Fisher, since the Royal Statistical Society refused to publish any more of his papers.

SEE ALSO:
▶ What to Expect, page 28
▶ Regression, page 118

Fallacies

THE MAIN JOB OF STATISTICS IS TO IMPROVE THE WAY WE DEAL WITH INFORMATION, replacing our natural, emotionally based predictions with correct ones, based on mathematics. But bad habits are hard to break and, compared to other sciences, statistics is still a new subject. So all sorts of fallacies (or misunderstandings) remain. Some, like "correlation means causation," have been covered already, but there are several more.

Right: The house always wins. Probability and statistics prove it.

Statisticians are not immune from making mistakes themselves, and one of these is the multiple testing fallacy, which means repeating a test and picking the answer they like best. For instance, someone might think that taller people are better at math, select 1,000 math students and work out the correlation coefficient (see page 124). Maybe this turns out to be 60 percent. The researcher should then conclude that there

is not enough evidence to prove the idea, and stop. But, he or she might be tempted to select another 1,000 students and try again. A different coefficient will almost certainly result, and there is a roughly 50 percent chance it will be higher. With enough repeats, any coefficient you like will eventually appear, but it will mean nothing at all. To see why not, remember that any test will have some chance of giving the wrong answer;

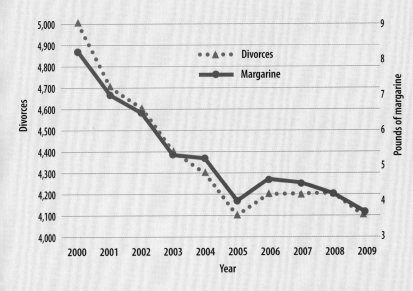

By searching through enough data, some meaningless correlations can be found, such as between margarine consumption and divorcee—the multiple testing fallacy.

are so strong that it's hard to believe it when they are wrong. A famous puzzle based on the TV show *Let's Make a Deal* is named after the host of that show, Monty Hall. (It is a slightly different version to the TV puzzle.) You are faced with three doors. Monty tells you that behind one is a million dollars while behind the others

maybe 5 percent. So, if you repeat it 20 times, it will probably give the wrong answer once. Picking that one occasion is cheating. This is why running both one- and two-tailed tests on the same data is cheating (page 169). After all, if changing the test changes the result from one you don't like to one you do, you might as well retry tests that do give good results and accept the second results, even if they are unwelcome.

is a rubber duck. You choose one of the doors (and tell Monty) but before you open it, he opens

Stick or bust?

Sometimes, our natural assumptions about statistics

Monty Hall used to suggest "Let's Make a Deal"—what would you do?

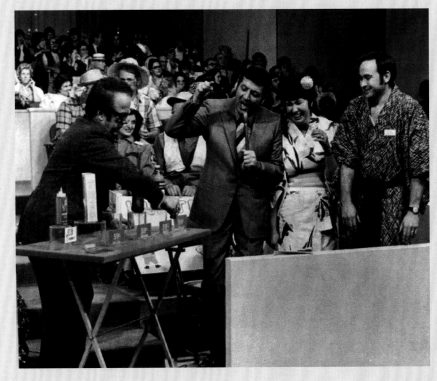

SMALL CHANCE, MANY REPEATS

Another kind of fallacy involves the difficulty in handling low probabilities. If there is a 1 in 1,000 chance of you getting hit by a car when crossing a road, would you feel safe to cross? Probably today, but what about next week or next year? If that road is outside your home, and you go out just once a day, the chances are you will be run over within a couple of years: 365 days x 2 crossings per day x 2 years = 1,460 crossings. And, if that probability applies to everyone who crosses that road (obviously a very dangerous one)—which might involve a hundred people a day—someone will be run over around once a week!

another, revealing a duck. He now asks whether you would like to change your mind about which door to open. Should you? Would changing your mind make a difference? It would. It will double your chances, in fact, if you change, you will probably win the million. If you don't, you are probably going home with just a new bath toy.

Million dollar math

As so often in the statistics of chance, the math here is very easy indeed. Here are the steps: there is a 100 percent chance that the money is behind one of the doors (let's call them A, B, and C). There is a one in three chance (about 33 percent) that the money is behind the first door you choose (let's call this door A). So the chance that the money is elsewhere (that is, behind B or C) is two in three (about 67 percent). Let's call the door that Monty opens first B. Since the prize isn't there, door B's chance of being the winning door is now 0 percent. None of this has changed the chance that the money is behind door A (still 33 percent), and since B is now 0 percent, the chance of C being the winning door is now 67 percent. So, switch and open it!

Lengthen the odds

To make this clearer, imagine a slightly different example. Now, the choice is between envelopes, just one of which has a check for a million dollars inside. Instead of just three choices, though, there are 1,000. To start with, you take one envelope. You're probably not that excited, and quite right, too, as there is only a 1 in 1,000 (0.1 percent) chance that the check is inside. Monty now opens

Door A   Door B   Door C

LOSE

Contestant's first pick   Monty Hall's revelation   Contestant's NEW pick

WIN

Contestant's NEW pick   Contestant's first pick   Monty Hall's revelation

WIN

Contestant's NEW pick   Monty Hall's revelation   Contestant's first pick

In the Monty Hall problem, switching choice boosts the chance of winning. The first choice is based on 1 in 3 odds, while the second choice (switch or not) has odds of 2 in 3.

all but one of the 999 remaining envelopes, and all 998 are empty. He offers you the final one. If you wish, you can switch to this final envelope or stick with the one you picked first. Should you swap? Since the chance of your first envelope being the winner is 1 in 1,000, the chance of the check being in one of the others is 999 in 1,000. Now that Monty has ruled out 998 of them, there is now a 99.9 percent chance that the prize is in that final envelope.

Birthday surprise

Another puzzle shows how difficult it is to get a feel for the way probabilities change. Imagine you're in a room full of people. How many people would the room contain to make it likely that two of them have the same birthday? 730? 365? 183? The clearest way to work this out is by listing the possible ways to *not* have the same birthday. Ann is alone in the room; the chance of her having a birthday this year is of course 365/365. Kofi joins her. The chance that he has the same birthday is 1/365, so the chance of Ann and Kofi having different birthdays is (365-1)/365 = 364/365. Clare arrives. Her chance of having the same birthday as Kofi is 1/365, and the chance of her having the same birthday as Ann is also 1/365, so the chance of Clare

Ann		Kofi		Clare		Derek		Maya		Frank	
365/365	×	364/365	×	363/365	×	362/365	×	361/365	×	360/365		= about 0.960

having a different birthday is (365 – 2)/365 = 363/365. So far, with three people, the chance of not having the same birthday is (365/365) x (364/365) x (363/365), which is about 0.992.

Ever decreasing chances

As we add more and more people, we keep multiplying by new fractions. So, for six people we have the calculation as shown at the top of the page. The chance of having a different birthday is getting smaller, and the calculation will get longer as more people arrive. To save space, we can plot a graph of how this probability changes as more people enter the room. The green line below is what we have been calculating, the chance of people not sharing a birthday. To find the chance of people sharing a birthday we just subtract that figure from 1, and this is shown by the red line. When we get to 23 people, we can see that the chance of not sharing a birthday is just under 0.5, and the chance of sharing is just over 0.5.

"So, if there are 23 people in a room, chances are that 2 of them share a birthday

Simpson's paradox

Simpson's paradox shows the fallacy of believing main, headline figures when more information is available. A medical study is looking at how well two different drug treatments work on small and large kidney stones, and this table summarizes the results.

	Number of stones treated	Number destroyed	Success rate
Alphazyme	349	270	0.77
Betachrone	352	286	0.81

So, Betachrone seems best. However, if we look behind the headlines to see the effects on small and large kidney stones separately, a very different story is revealed (See the larger chart top, right). So now we see that Alphazyme

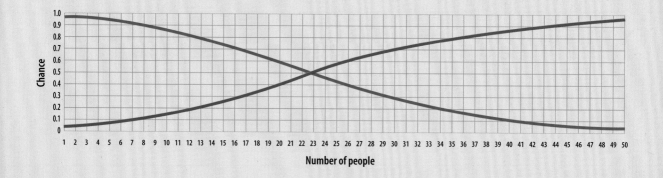

	Number of stones treated	Number destroyed	Success rate
Alphazyme, effect on small kidney stones	89	80	0.90
Alphazyme, effect on large kidney stones	260	190	0.73
Alphazyme totals	**349**	**270**	**0.77**
Betachrone, effect on small kidney stones	270	230	0.85
Betachrone, effect on large kidney stones	82	56	0.68
Betachrone totals	**352**	**286**	**0.81**

THE GAMBLER'S FALLACY

The gambler's fallacy is that objects have memories. Although this doesn't sound very likely, it's a fallacy that most people are tempted to believe. It has ruined many gamblers, who should know a thing or two about chance, especially at roulette, which uses a wheel with equal numbers of black and red spaces. At the casino in Monte Carlo, (below), on August 18, 1913, the roulette ball had fallen into a black hole 25 times in a row. Surely, people thought, the next hole it falls into must be red? Millions of francs were bet on this outcome, and all were lost when the next hole was black, too. Of course, getting 26 blacks (or 26 reds) in a row is very unlikely, with odds of about 1 in 6.6 million. But the gamblers were betting on getting just one red, and the odds of that are just 1 in 2.

HOW STRANGE?

Every so often, the most amazing coincidences happen, like thinking about someone you've not heard from for a long time and then getting a call from them the next day, or coming across a new word and hearing it again a few days later. But how amazing are these situations? What really is amazing is how many things we experience, and how many thoughts we have, every day. We might easily hear and read a million words a day and think a million thoughts, too, and each week we probably think about almost everyone we know, only to forget nearly all these thoughts straightaway. So if anyone we know calls, we've almost certainly thought of them recently. And, when it comes to the words we see and hear, we rarely notice them at all. We aren't interested in them as individual things, only in what groups of them mean. But, we are exceptionally good at spotting things that interest us (like our own name, even when very quietly spoken in a crowd). So, if, just for a change, there is a particular word that we have noticed because it's new, we will spot it at once and ignore as usual all the thousands of others we hear.

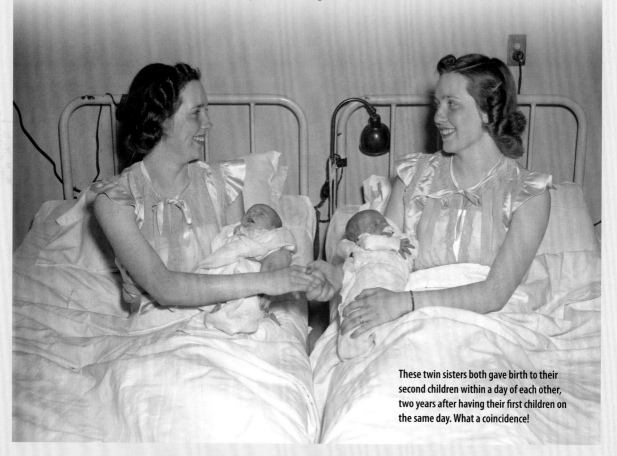

These twin sisters both gave birth to their second children within a day of each other, two years after having their first children on the same day. What a coincidence!

works best on both large stones and small stones, and only when we ignore stone size does Betachrone do better.

Lurking variables

This strange result can happen when we lump together things that have some hidden difference (sometimes called a lurking variable). Here, the difference is that larger stones are harder to treat. Doctors know that Alphazyme works best, but it has nasty side effects. So, they use it mainly on large stones. Although Betachrone is less effective, it is used more frequently on smaller stones. Because these are both more common and easier to treat, Betachrone has more successes overall.

The power of coincidence

Many statistical fallacies are based on confusing two very different numbers:

A. The probability of something happening
B. The probability of something happening to you.

If the "something happening" is "winning the lottery" then these numbers are different. A is 1, (or 100 percent), and B is less than 0.0000001. In words, this confusion is wrapped up in the statement "Almost every week, someone wins millions on the lottery, and it only costs a dollar, so it's worth doing." The point is that the chance of that person being you is extremely remote. So, most likely, you will lose dollar after dollar.

P.T. BARNUM & CO'S GREATEST SHOW ON EARTH & THE GREAT LONDON CIRCUS COMBINED WITH
SANGERS ROYAL BRITISH MENAGERIE & GRAND INTERNATIONAL SHOWS

HOROSCOPES AND THE BARNUM EFFECT

Does this sound like you? You are a fairly easy-going person who is quite often misunderstood. You can work very hard if necessary, though you usually prefer to avoid tasks that are too demanding. You are often too critical of yourself. You are an independent thinker, and are unwilling to accept other people's opinions unless they can provide strong arguments or proof that they are right. You can be impatient, especially when other people are indecisive. You have a great deal of untapped potential. If it does sound like you, then you may be experiencing the Barnum effect (named after showman and skilled manipulator Phineas T. Barnum—see more about him, page 110). The Barnum effect is the natural tendency to exaggerate the significance of what is true and ignore what isn't true. This is why horoscopes can seem spookily accurate. To see whether they really are, researchers jumbled the star signs before the readers saw them. When they did, people tended to think that the horoscope labeled with their own sign was the most accurate, even though it had actually been written for someone else.

SEE ALSO:
▶ The Normal Distribution, page 46
▶ Randomness, page 112

Who Makes Your Decisions?

STATISTICS PLAYS A MUCH MORE IMPORTANT PART IN ALL OUR LIVES NOW THAN IT DID 20 YEARS AGO, because there is enormously more information about us available online, and also because computer processing power is so quick and cheap that analyzing all that data is easy. For instance, for the first time in history, shops and sellers know more about their customers than the customers do themselves.

Since 2004, the huge grocery company Walmart always has its stores stocked with extra beer and pop tarts when a hurricane is on its way, because its owners know (because statistics tells them so) that these products sell very well before and after hurricanes arrive. Often, this new statistical power is very useful: if you love tennis while your friend loves gardening, you will get different results by typing "lawn" into a search engine, because that search engine has recorded and analyzed what you have searched for in the past and used that information to build up a profile of you and your interests.

When you next come across a bargain or treat in the store, think about how the managers already knew you would want it.

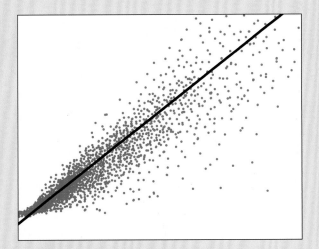

Almost all the tools used by modern automated statistical programs are the same as have been used for decades. The most important is regression (see page 118), which fits a straight line to a scattering of data. It gives highly reliable results thanks to the huge datasets now available.

Experimenting on everyone

The statistical techniques for tests and trials, like confidence limits and the t-test, are also far more powerful now that they can be used to experiment on huge numbers of people. For instance, rather than trying to work out how much interest their customers would pay for a loan, insurance companies will sometimes send a number of offers to all of their customers, perhaps offering 10,000 customers a 2 percent interest rate, another 10,000 a 2.1 percent rate, and so on. But this isn't used to find out the most acceptable rate and charge everyone that. Instead, those who are willing to pay more than others are charged more, whatever the reason they pay. They might be careless with money, or not understand the details, or assume that that is the rate that everyone has to pay.

Emotional connection

Would you be happy to pay more than anyone else for something, just because the provider knows you are more likely to pay that amount when others may walk away? Experiments like this also exploit the fact that our emotions are involved in decisions even when we don't realize it. For instance, simply including a photograph of a person on an offer letter will increase the chances of the offer being accepted.

The end of experts?

There's nothing new in appealing to people's emotions, of course. Puppies have been used to sell toilet rolls for decades. What has changed is that the power of this kind of advertising is now far greater than the best human salesperson. One reason this has changed the world is that human experts are really not very good at understanding and predicting the behavior of other humans. And they—like everyone else—tend to be enormously more confident that they are right than they should be. If you ask someone about almost anything: "What is the population of your town; what will the weather be on your birthday; how many minutes did you spend on the phone yesterday?" they will usually give a rough answer. Then ask them to put a confidence interval on that answer (see more, page 154) so that instead of saying "about 20 minutes on the phone," they say "between 15 and 25 minutes." Those ranges

BIG DATA WORLD

The term "Big Data" refers to the vast amount of information available about all sorts of things online. The idea is linked to the Internet of Things. While the early Web connected computers, today it connects people (through social media and the like), and in future the Web will link up all kinds of machines, sensors, and everyday objects like cars, fridges, and houses. The Internet of Things provides Big Data, which is distinct not just because of the amount of this information that is new, but also because it is different to the kind available in the 20th century, since it includes data from so many sources. It is possible to gather vast amounts of weather data, health information, and commercial activity and analyze it all along with things like traffic figures and water use. That analysis will reveal unsuspected links between phenomena, thus offering a chance to predict and preempt problems with confidence. And, of course, the information will help with marketing products. Personal data is delivered more or less live, which is why you might one day soon receive personalized shopping offers as you walk past a store.

will often be very narrow, with the true answer well outside them.

Computers know best

Sometimes it is obvious that trial and error will give better results than experts. The way people respond to adverts on a computer screen varies according to the position, size, color, of wording and to the use of images, video clips—and to who is using the computer, when, where, why and how. With so many variables, it would be a brave psychologist who would even try to predict which adverts will be most

effective. But, if the screen is showing a popular website or search engine, all that is needed is to try several kinds of advert out every hour and the most successful one will be revealed simply by

Advertising works, but statistically driven marketing works better!

NO MORE PRIVACY

In many countries these days, when people apply for jobs, training courses, apprenticeships, or university places, their interviewers will often know all about what that person has said on social media, even decades earlier. And, if they get the job, that person's emails will often be monitored. Sometimes, even time spent in the restroom or moving between desks is tracked. Some companies give their employees free phones, but the hidden cost may be that all the calls and texts on that phone will be monitored. All this information can then be used to decide whether to promote, reward, or dismiss that person, based partly on statistical analyses which will predict their future value and risk to the company. Where will this technology be used next?

seeing who clicks on it. Since popular pages are viewed by millions of people a day, the results are likely to be very reliable indeed.

Statistical takeover

Statistics is a great source of knowledge about past, present, and future, and its power is growing fast. It is by far the most successful way of understanding human behavior. Vast amounts of data about each of us, together with many of the statistical tools in the book, are combined to predict and influence what we think and

what we do. This means that, like most human inventions, statistics can be used for good or for ill. But understanding how statistics works and how it can be used and abused teaches us how to understand the dangers, and how to avoid them, too. And, it also shows us how to find out about the world, make good decisions, and find all the certainty there is.

> SEE ALSO:
> ▶ Bayes' Amazing Theorem, page 52
> ▶ The Average Human, page 96

Glossary

Census

A survey that collects data from every member of a population. Often, this is so difficult or costly that a sample must be used instead.

Correlation

Correlation coefficients measure the strength of a relationship between two things, such as height and weight. In a linear relationship, if you plot a graph of one thing against another, the points you plot will tend to form a straight line. A positive correlation means that if one thing gets larger, the other gets larger too, and the line will slope upward. A negative correlation means that if one thing gets larger, the other gets smaller, and the line will slope downward. The value of a correlation coefficient lies between -1 and 1.

Degrees of freedom

If a sample of 50 items is available, and 2 population parameters are estimated from it (such as mean and standard deviation) then the number of degrees of freedom is 50 − 2 = 48. Several tests and statistical formulas make use of this number.

Distribution

A list of all the values of a group of things, together with how often each value occurs.

Expected value

The average value of a random outcome that has been repeated many times. So, the expected value of a dice throw is 3.5.

Hypothesis

A theory, suggestion, or assumption about a parameter.

Least squares

A method for checking whether a line is the best fit to some points. If it is, the sum of the squares of the distances from the points to the line is as small as it can be.

Mean

The result of adding up a group of values and dividing by the number of values. This is what people usually call the average.

Median

The middle value in a group, which divides the group in half.

Mode

The value in a group that occurs most often. Some groups may have no mode (like 1, 4, 5, 7, 9) or more than one mode (like 1, 2, 2, 3, 6, 7, 7, 8, 9, 9).

Normal distribution

A distribution that is symmetrical (that is, the left and right halves are mirror images of each other) with the mean, median, and mode all at the same place. Sometimes called the Gaussian distribution after mathematician Carl Friedrich Gauss.

Odds

The chance that something will happen. If the odds of some event taking place are 3:5 ("three to five"), then we expect that the event will happen 3 times for every 5 times it does not occur.

Outlier
A value which is suspicious because it lies outside a group of data.

Parameter
A number that helps to define a population, such as mean or standard deviation.

Population
The entire group of items from which data can be selected, such as all the people in the world, all the stars in the Universe or all the light bulbs made by a factory.

Probability
The chance that a particular thing will happen, stated on a scale either from 0 (impossibility) to 1 (certainty) or from 0 percent to 100 percent.

Random
A random sequence (list) of numbers has no pattern and cannot be predicted.

Regression
A method of finding the equation of a line that best fits some data. The simplest type is linear regression, which finds the equation of a straight line, usually by the least squares technique.

Risk
The probability of an event occurring, especially when the event is unwanted.

Sample
A part of a population that is collected (usually randomly) and then used to study the whole population.

Sample size
The number of items in a sample.

Significance
If a relationship between two things is significant, then it is not due to chance.

Skew (or skewness)
Lack of symmetry. If a distribution is negatively skewed its left tail is longer. If it is positively skewed then its right tail is longer.

Standard deviation
A measure of the spread of a group of data. The larger the standard deviation, the more spread the data is.

Variance
A measure of the spread of a group of data. Variance is equal to the square of the standard deviation.

Z-score
The gap between a value and the mean, measured in numbers of standard deviations. So, if the mean is 6, the standard deviation is 1.5, and the value is 9, the Z-score is 2.

• Many of the example calculations in this book rely on rounded data, so the results are approximate values.

Index

Cataloging-in-Publication Data has been applied for and may be obtained from the Library of Congress.

ISBN 978-1-62795-144-9

Design: Bradbury & Williams
Copy editor: Meredith MacArdle
Proofreader: Julia Adams
Consultant: Kevin Adams
Picture Research: Clare Newman
Cover Design: Wildpixel Ltd

SHELTER HARBOR PRESS

603 West 115th Street Suite 163
New York, New York 10025
www.shelterharborpress.com

For sales, please contact
info@shelterharborpress.com

Printed in Shenzhen, China.

10 9 8 7 6 5 4 3 2 1

PICTURE CREDITS